Tinkering Tall Ships

A memory book from
Doc "O"

www.DocOsBooks.com

www.DocOsBooks.com

Printed in USA

Published by:
SAN 299-2701
Glenhaven Publications
4262 NE 125 Street
Seattle, Washington 98125

ISBN 978-0-9657061-2-4 (paperback)

Table of Contents

Introduction
Ways
Slipway
Blocks
Shores
Keel
Ribs
Keelson
Cross bands
Ribbands
Cradle
Bow
Stern
Tumblehome
Poop deck
Fo'c's'le = forecastle
Step a mast 1/5 1/2 3/4
Bowsprit
Beakhead
Seat of ease
Stays vs. shrouds
Bob stay
Bowsprit shrouds
Cap
Jibboom
Dolphin striker
Martingale
Transom
Rudder
Bulwark, Railings
Wales (gunwales= gunnels)
Channels = chain wales
Hounds
Cheeks
Trestle trees
Cross trees
Shrouds 4, 5, 3
Backstays
Forestay
Top
Lubber's hole
Cap
Top mast
Fid
Doublings
Topmast cheeks
Topmast shrouds

Tinkering Tall Ships

Futtock shrouds=foot hooks
Topmast stays
Yard-arms
Lifts
Slings
Cranse-irons
Braces
Sails

Rig check day
Event day in the gym
Yummy Tummy

Tinkering Tall Ships

Introduction

Tinkering.
That's a verb. Sort of.
It is something you do.
For us, it is something you do to help you understand an idea.
Tinkering helps you see the problems. It helps you notice what you don't understand.
Yet.
Yet is one of my favorite words. It means I am going to learn more. I'm in the midst of a process. I'm not finished yet.

The purpose of the Engineering Event is to provide the opportunity to experience the way engineers think.

1. Design something to do a task.
2. Fix it when it breaks.

To engineer is to make a working mechanism.
We make ours of junk. Cardboard and plumbing parts and lots of glue and tape.
Children make models to comprehend the structures – not to make accurate models.
The design starts with the task to be done.
How could we make a device to move like a car? An umbrella? A grappling hook?
Then breakage is anticipated. The ability to make easy repairs is made part of the design.
Finally, the outside is made pleasing to look at.

We teach children.
They focus on objects and object constancy.
Children play. They like to make things. And make them better. And be cheered for success.
Children are social beings. They will work for praise and appreciation.

We start with a theme. Something in the news. This year, Tall Ships were in the news as the Great Lakes festival had to be moved from Cleveland to accommodate the Republican National Convention. But what did we know about Tall Ships last summer? Not much.

Concepts we explored this year.

1. Hulls are made to do the task of carrying goods. Great big delivery trucks.
2. Hulls have to be strong and durable. We have to launch them. So we need cradles.
3. Sails allow wind to deliver the hull. Fast.
4. Masts break. Sails are stressed so much they are ruined.
   How do you fix the system so you can still go fast?

The Web is a wonderful place to study topics such as this. The 1929 essay on *The Development of the Clipper Ship* by Charles E. Park from the American Antiquarian Society, gives a delightful introduction. A great book of clear and detailed perspective drawing sketches of ship rigging is *Rigging Period Ship Models* by Lennarth Petersson. (First published by Chatham Publishing, 2000) Another useful reference is *Ship Modeling Simplified* by Frank Mastini (International Marine/ Ragged Mountain Press, 1990). The most interesting and useful reference we found, full of historical anecdotes is *The Ship Model Builder's Assistant* by Charles G. Davis. (Dover Publications, 1988. Originally Publication Number Twelve of the Marine Research Society, Salem, Massachusetts, 1926)

The faculty of our school asked if these were slave ships. The answer is "No". An ABC news article of July 2, 2016 explains that the large-scale African slave trade was between 1640 and 1680. Our ships were 200 years later. In 1794 the U.S. Congress passed laws "prohibiting the manufacture, fitting, equipping, loading or dispatching of any vessel to be employed in the slave trade." You got arrested if you had slaves on board. You got arrested if you had a "slave deck" - even if you never used it to transport slaves. The deck had to be removed. Clipper ships were built between 1843 and 1855, mostly in Boston by famed Naval Architect Donald McKay. They were designed to make money taking people and food to the gold fields of California and Australia. On the way back they went around the world to China and got a load of tea. The California gold rush happened in 1849 and the Australian gold rush happened in 1851. That is about 40 years after slave ships were banned from the seas. Clipper ships were fast and they made a lot of money. Construction of Clipper ships abruptly ended with the arrival of the steam engine. Steam moved ships faster and didn't rely on changing wind patterns.

TAXES DETERMINED HULL SHAPE

Ships sail into harbors to deliver goods.
But harbors are owned by countries.
Governments realized they could get tax revenue by taxing ships unloading in their harbor.
Taxes were common in China and India but the first European tax seems to have been levied by Kind Edward I in 1202. The tax was based on how much the ship could carry. It was called BURTHEN in Early Modern English and BURDEN in Modern English. Many convoluted formulas were devised to measure Tons Burden.
Example: Length –(beam x 3/5) x beam x beam/2 all divided by 94

In 1377 the tax was 3 shillings per ton of wine
In 1678 it was measured by 3/5 of the ship's displacement.
By 1694 it was based on LxBxD divided by 94
Finally, in 1836 a new, simpler calculation was made into law. It was decreed that all measurements are taken from the ship's midpoint only. Depth was from the floor of the hold to the uppermost full deck. This excluded the keel. Aha! If you made more decks that didn't go the full length of the ship, they were tax free.

To avoid taxes, hull shape changed.
Bows became more squared off to gain free cargo space.
Hulls bulged out sideways below the midpoint and above it.
Tumblehome.
More and more decks on the bow and stern were added, as well as "houses" in between
Forecastle. Poop deck.
Hulls were also shaped specifically to carry boxes of tea. They could earn a lot of money on a single trip. Hulls could hold 40,000 chests of tea. Each chest weighed 140 lbs and sold for $28. That means each tea delivery earned more than a million dollars. Enough to pay for the entire construction of the ship in one trip.
Each Chief Mate was given 3 chests to sell and each crew member had 2.
Land-lubber jobs paid bout $30 per year so a sailor could earn a entire year's wage on one sailing trip. Sailing was hard work but everyone earned a lot of money.

Ships were expected to last about ten years in the rough punishment of the seas. After that the remaining hulk was used on our west coast and in Australia to haul lumber and salmon until it confronted a disaster and sank.

Machines take fuel. You have to put gas in your car. Humans were the machines that worked the ropes and sails on the ships. They needed fuel. Food. By today's standards, the diet of the sailors seemed pretty bad. Beef jerky, stale buns, sauerkraut, and rum. Actually, that was probably a better diet than many would have on land. At least they had protein every day. And captains searched for ways to keep the crew healthy. Sauerkraut provided vitamin C to stave off the dread disease Scurvy.

The National Maritime Museum in Greenwich, London stores records of how the Royal Navy recruited boys to join the navy. Roland Pietsch wrote a doctoral thesis on "Ships' Boys and Charity in the Mid-Eighteenth Century." (University of London, 2003) He says boys typically joined the navy as trainee sailors at age 13 but some officer's sons joined at age 11. They were short. About four foot three when they joined and only five feet by age nineteen. They began earning full wages by the age of eighteen. Some became ship captains by age nineteen.

Nathaniel B. Palmer (Palmer station, Antarctica, clipper ship N.B. Palmer, icebreaker RV Nathaniel B. Palmer) was such a boy. Born in Stonington, Connecticut, he grew up dreaming of sailing. Sailing fast! According to Basil Heatter *in Eighty Days to Hong Kong: The Story of the Clipper Ship*, before the age of twenty, Palmer was in command of the coastal schooner Galena. He then joined the crew of a seal hunting ship, the Hersilia, bound for islands beyond South America. He is credited with first sighting land to the south of there. Thus he was the first sailor after Columbus to discover a new continent: Antarctica. The Palmer Peninsula bears his name. As he changed command from ship to ship over the next several years, he tried to understand how to make ships go faster. He found that a flat bottom ship carrying bales of cotton was surprisingly fast. Most people thought fast ships had v-shaped hulls. But not Palmer. He began working on the design of a long, narrow, but flat bottomed ship that would carry large loads and also be fast. Palmer met Edward Collins who owned a large shipping company. They talked about design. They then found John Willis Griffiths, a brilliant young navel architect who agreed with their design ideas. Finally they located a young shipbuilder who agreed to build their plan. He was virtually unknown and so had no reputation to lose if the ship didn't work. The young builder was Donald McKay, who went on to become the most famous of them all. These four young men, Palmer, Collins, Griffiths, and McKay developed the dream of huge, fast ships, building several ships per year at McKay's shipyard. McKay tinkered and planned larger and larger, faster and faster ships such as the world had never seen. Clipper Ships. Many New England ship builders joined the excitement of new hull designs. In 1851, thirty-one Clipper ships were launched. Thirty-three Clippers launched in 1852. 1853 was the pinnacle of Clipper ship building with forty-eight launchings. By 1854 the number declined to twenty..

The largest Clipper ship ever built, the Great Republic, was launched by McKay in 1853. It was 335' long, 51' beam and had four decks and four masts. The Great Republic was expected to be the swiftest and finest Clipper ever built. It was taken to New York harbor and filled with wheat to be delivered to gold miners in Australia. The sails were furled and it was ready to depart when a fire broke out in a bakery on the wharf. Bucket brigade fire crews were unable to control the blaze. Sails and rigging caught fire and the crew tried to save her by cutting the rigging. The fore-topmast broke in two and stabbed down through three levels of decks. The fire crews left thinking the fire was out. Two days later it was found that the wheat cargo was burning. They scuttled the ship to put out the flames. They later found the seawater had caused the wheat to swell straining and buckling the ribs and knees of the hull. After a year and a half of repairs and reducing the size of the ship to three masts, she sailed out and crossed the Atlantic in thirteen days, one hour. Still the fastest ship in the world.

Adventure. And inventing. And improving the design.
That is what Clipper Ships were about.
That is what our tinkering project is about.
Not unfocused tinkering, but tinkering to come to understand the ideas we are thinking about. Learning about a new field of endeavor through crafting junk into models of it. Tinkering with junk helps us process our understanding and store that in long-term memory.

Tinkering and tool use. Glue guns, pliers, tape, string, needle-nose plier nippers, marking pens, scissors, hack saw, vices, and a drill press. Cardboard boxes, paper soup lids, berry baskets, toothpicks, foot-long dowels of various diameters, tongue depressors, drink straws, wheels and axles, rubber bands, paperclips, binder clips, large coffee filters, and brass washers.

This is a book of wacky, weird, wonderful ideas. Tinkering ideas. But mostly this is a memory book for the children.

This book may help us;

1. Share ideas as we tell others about our adventure.
2. Remind ourselves about how our ideas developed.
   To remember when we didn't know.
3. Celebrate moments when we were clever. When we were capable. Confident.
   A glue gun user. A hacksaw user. A drill-press user. A competent tinkerer.

Tinkering

Tall Ships

A memory of our work together.

Tinkering Tall Ships

It was October. Doc "O" was going to announce the theme for this year's Engineering Event. Students could hardly wait.
The spaces for letters were drawn on the board.

____ ____ ____ ____   ____ ____ ____ ____ ____

Now, what could that be? Someone guessed a letter. And so did others as the pointer moved around the class.
TALL SHIPS
That is an odd title.
Why did we choose that? It is supposed to be something in the news.

The Republican National Convention brought Tall Ship news.
Cleveland was supposed to have a Tall Ships celebration. From there the ships would tour around the Great Lakes all summer. It turns out there weren't enough hotels in Cleveland to have two celebrations at once so Tall Ships had to move to Fairport Harbor about 30 miles away.

There is a photo of a Tall Ship on the wall. It sure is pretty. And complicated.

These ships were delivery trucks.
From country to country across the seas.
A long time ago China wanted to sell stuff throughout the world.
Christopher Columbus was looking for new trade routes.

Ocean sailing happened for centuries.
1421, 1492, 1849, 1851.
China. Columbus. The California Gold rush. The Australian Gold rush.

Most of the really tall ships were built in the U.S. where ship builders didn't have strict government rules.
They were called Clipper Ships because they sailed at a fast clip.
Records of 16mph, 18mph and even 20mph were recorded.
That was fast for world travel in those days.

How big were they?
Mostly from 150 feet to 200 feet long.

Boston naval architect Donald McKay built many of them.
His ships ranged from150' to 250' in length.
Only six famous American clipper ships were longer:

One of the first really tall ships was built in Boston named the Sea Witch.
The longest clipper ship ever built was 302'.
It was the Great Republic by Donald McKay.
Of 370 American clipper ships, only 30 were 210' or longer.
The most common length was 185'

How wide were they? Width is called BEAM.
How deep were they? Depth is from keel to topmost full deck.
We began listing ship sizes. We made a chart.

Name	length	beam	depth
Stornoway	157.5'	25.7'	17.7'
Race Horse	125'	30'	16'
Challenger	174'	32'	20'
Sword Fish	169.5'	36.5'	20'
Sea Witch	170'	34'	19'
Great Republic	302'	53'	39'

So the ratio of length to width was from 4 to 6, mostly 5 or 6.
The ratio of depth to length was from 7 to 9, mostly about 8.

If we want to make hull models about 2' long, they would be 4.5" wide.
They would be about 2.5" to 3" deep.
Students were getting excited.
They were ready to start.

What is the first question?

Where?
The first question is ... where?
Before we start to build, we have to ask where?
WAYS.
Slipways.
A flat parking lot-like space near a river or body of water.
Then, how do we hold up the hull as we build it?

Of course then we need to support them.
Ships were built on WAYS with blocking and cradles.
At launch time the cradles were wedged up to support the weight of the ship and the blocks were knocked away The hull slid down the way.

Students looked around their houses for something to hold a ship.
An egg carton in the garbage!
The very thing!

All these lumps could prop up a ship.
Egg cartons were set out and we were ready to begin.
We need a keel.
Ships begin by laying a keel.
What can we use?
We could buy sheetrock corners, but they have metal which could snag skin.
Isn't there some plastic sheetrock corner? Nope.
OK then. We can try to remove one side from a J-channel.
How can we "peel" a J-channel?
Pliers held at right angles and twisted a bit are able to remove the strip.
But first we have to learn about pliers.
We refer to humans as tool users but most kids in my class do not use tools.
They play pat-a-cake, They tap keyboards. They do not know how to use their hands for strength.

So, here is a pliers. It is not a scissors. It is two tools in one. Open the handles. Move the sides back and forth in the slot. One way lets the jaws close. The other way keeps a little space between the jaws. We need to grip the J-channel firmly so be sure your pliers are adjusted the right way. But we still need a strong grip. Many of us have played baseball. We know about "choking" the bat. That gives more control but less strength. Our natural tendency is to "choke" the pliers. Moving your hands down to mid-handle gives more strength to the grip.

Now peel. Doc "O" says you will have to nibble about 60 times to clear the two-foot length.
Amazingly, we did it! It took a long time to get the cadence figured out, but we did it.
We peeled the whole keel.

Set the keel in the egg-carton blocking.

Now, how can we make ribs?
What material can we use?
A small cardboard platter might work.
It is difficult to cut out.
The half-circle ribs fall over.
The biggest problem is the size. Nine inches across.
Ships with ratio of 1:6 would be 54 inches long.
With so many students, we would have 200 ships over four-feet long in this classroom. That would be overwhelming.
Are there smaller platters? I can't find any smaller than 6-inches.
Hummm.
Oh! Maybe soup lids. They are about 4-inches in diameter.
And they stand up on their edge.
Cut one in half. You have 4-inches wide and a two–inch depth.

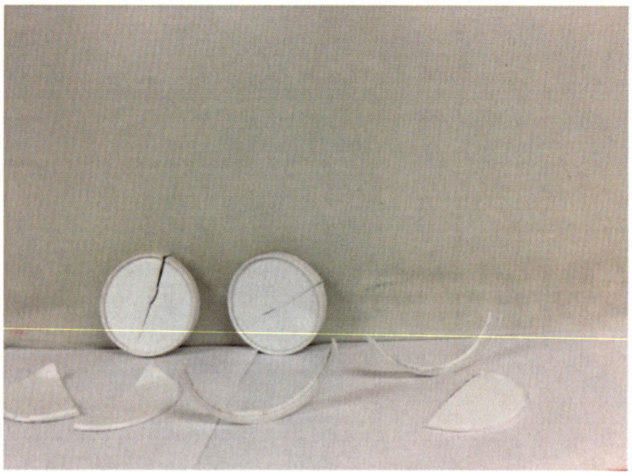

That is just about right.

Looking carefully we can see that the membrane is not glued to the soup lid.
It is just crimped in.
If we push firmly we can remove the membrane to get a rib.

How many will we need? Our keel is 24 inches.
If we place a rib every two inches, leaving a one inch space at the ends, we will need 12 ribs. How many soup lids? Six.
Draw a line down the center and cut them in half.

These ribs have a tendency to relax and open wider over time.
Looking carefully at pictures of ship ribs under construction, we see each rib is set up with a cross-band.
Our ships will be 4-inches wide. We can make 4-inch cross bands out of masking tape.

Now where shall we place them on the keel?
Make black pen marks on the keel every two inches.

Put a dab of hot glue on a pair of marks and set the rib in place.

We were careful to line up the crossbands to make a level ship deck.

Cradles were placed under ships to launch them.
Ships slid down the ways on cradles.
If we plan to set the rib-hull on a wheeled cradle and slam it across the classroom floor, the hull needs to be strong. Clipper hulls were strong. They withstood ocean waves.
Looking at Clipper designs, we notice keelsons were added for strength.
We have some old 1/8 inch strips of Balsa wood. They are two-feet long.
Just right! And quite light weight. So that might be easy to move with a rubber band.
Now, insert the Balsa-wood keelson and dribble glue down on it over each rib.

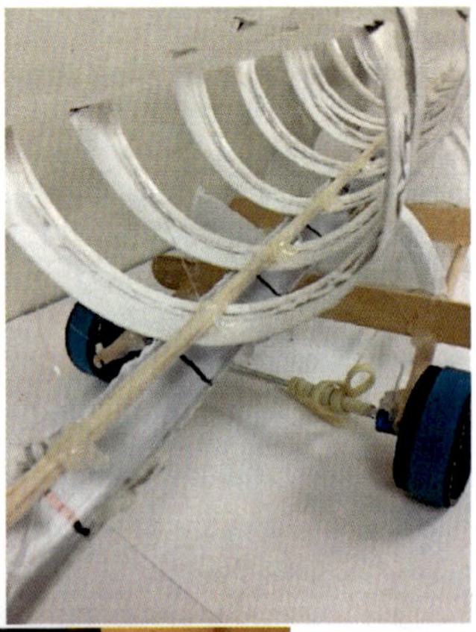

Done!. That looks pretty good.

Many people decided to add masking-tape ribbands to strengthen the hull.

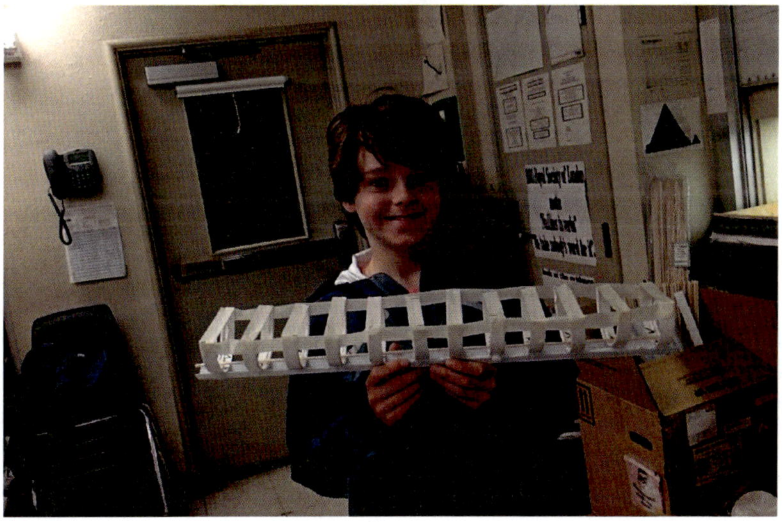

Now for the cradle.

What can we use for a cradle frame?

We can cut 36-inch dowels in half making an 18-inch cradle.

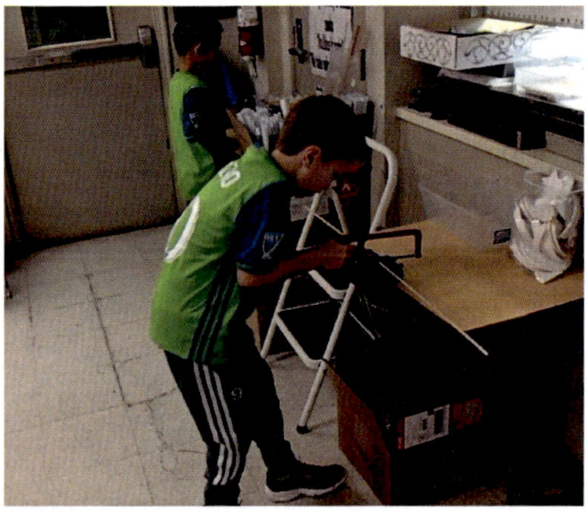

Everyone loves to saw. It is fun. But you have to measure carefully or the two pieces are not the same length.

How can we make cradles to carry the rib-hulls?
We need a V-shape to conform to the hull.

We have lots of left over membranes torn from making the ribs.
We notice fat tongue depressors. Six-inches long.
Our hulls are 4-inches wide so that is just about right.
Mark one inch in from each end leaving 4-inches in the middle.
Cut wedge shapes from the left-over membranes.
Hot-glue two membranes on a tongue depressor.
How many do we need?
Probably five would work pretty well.

Glue five tongue depressor cradle sections on the dowels.

We have cradles.
But we want to race.

We need wheels.
Pitsco wheels always work pretty well for us.
We need the axle to be about 6-inches wide to span the ship.
Don't forget the straw bearings so the wheels will turn.

You have to push hard on the pliers to get the axle to go in.

We need power to race it.
Contest rubber makes a fine motive force.
Potential and Kinetic energy.
It looks like we need another dowel to attach the contest rubber band.
Make it 6-inches long, just like the axles.

So now, how can we make it move?
We are having trouble learning how to snaggle the rubber band around the axle.
Then we have to turn the wheels to wind it and stretch it.

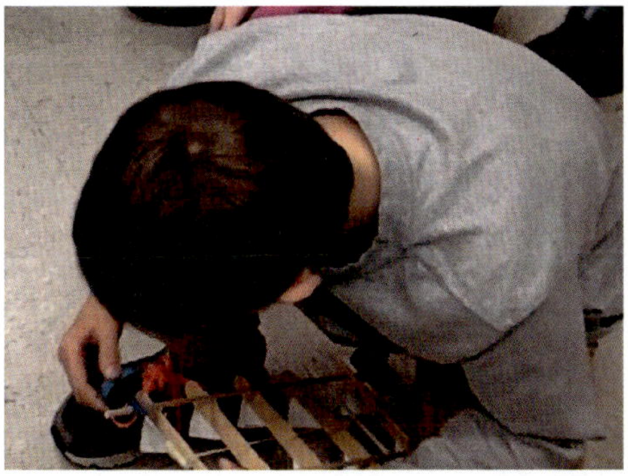

Stretch the rubber band
while you are winding.

Just winding the
rubberband
around the axle
doesn't make it move.

Notice the shoeboxes. Each child has one. Small parts and in-process items are stored there. Each child has a 3' long shelf to store their shoebox, cradle and rib-rig. They have another storage rack for their tall ship.

It is against the rules to get into other's stuff. Your stuff is safe. You don't have to worry or protect it. Physically safe. Emotionally safe. Free to tinker.

Now,
hold the wheel
tight as you set
the rig on the
floor

It keeps slipping!

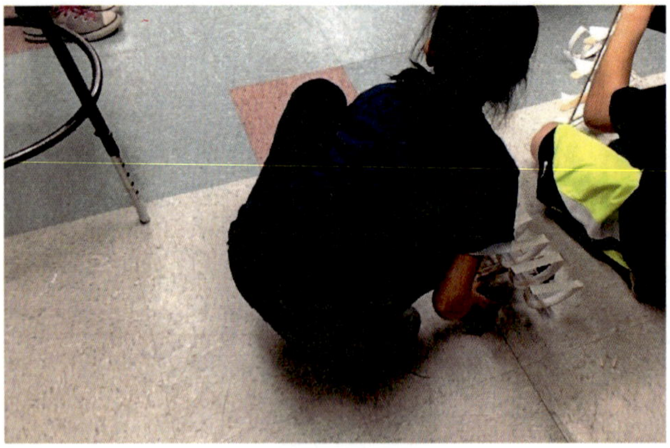

Wind it carefully without the rib-rig.

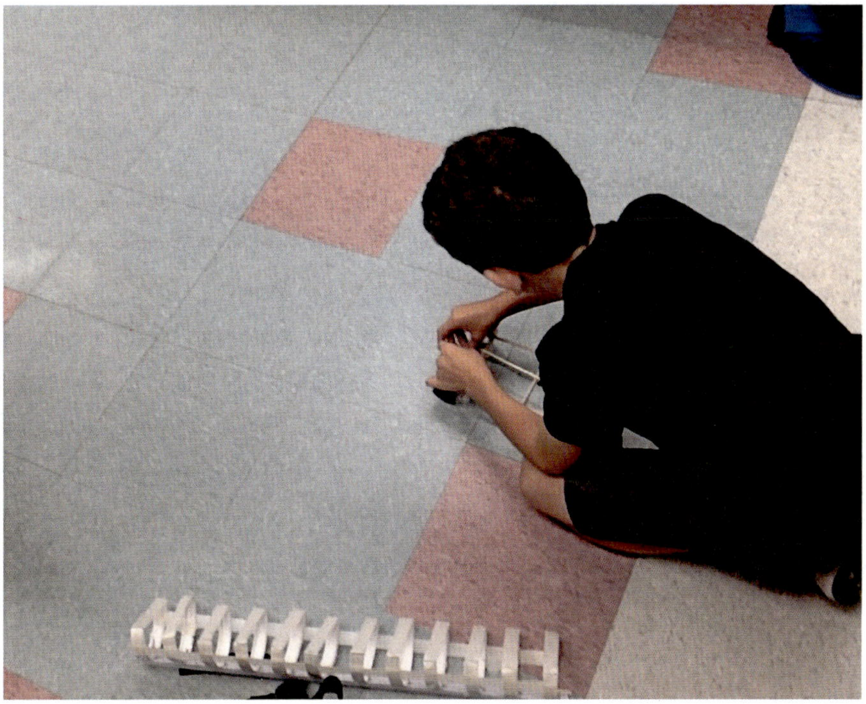

Everyone starts to figure it out.

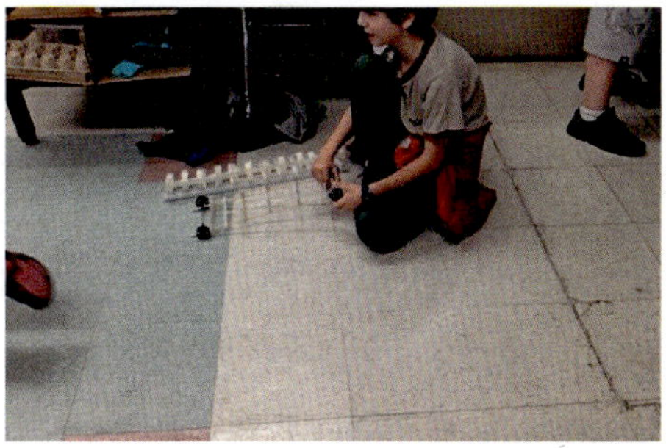

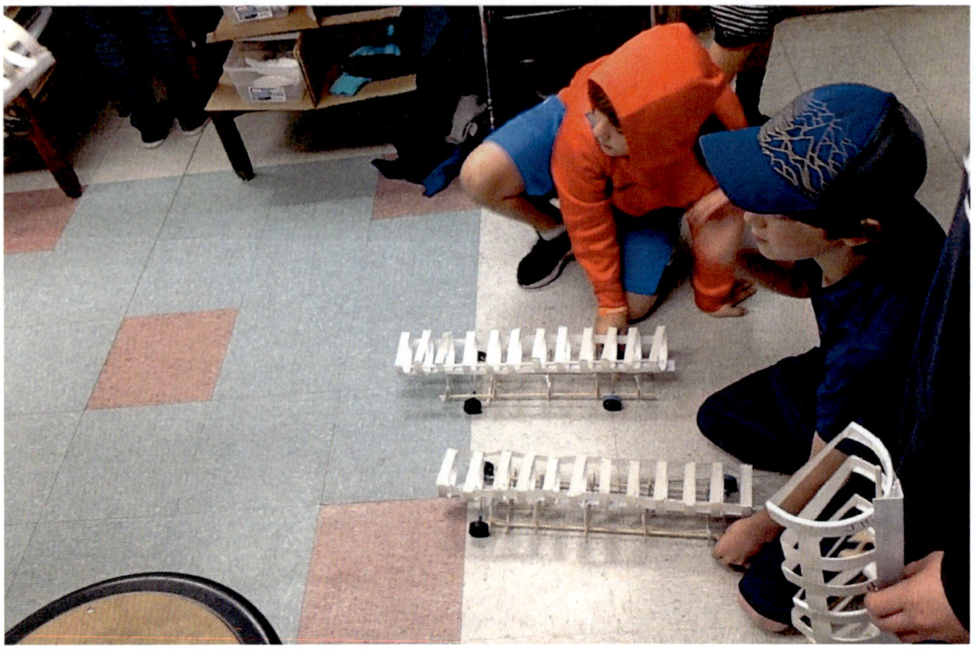

"Clear the Course!"

"Witness, Witness!"

Wind carefully.

Pay attention to the coil.

Make it neat.

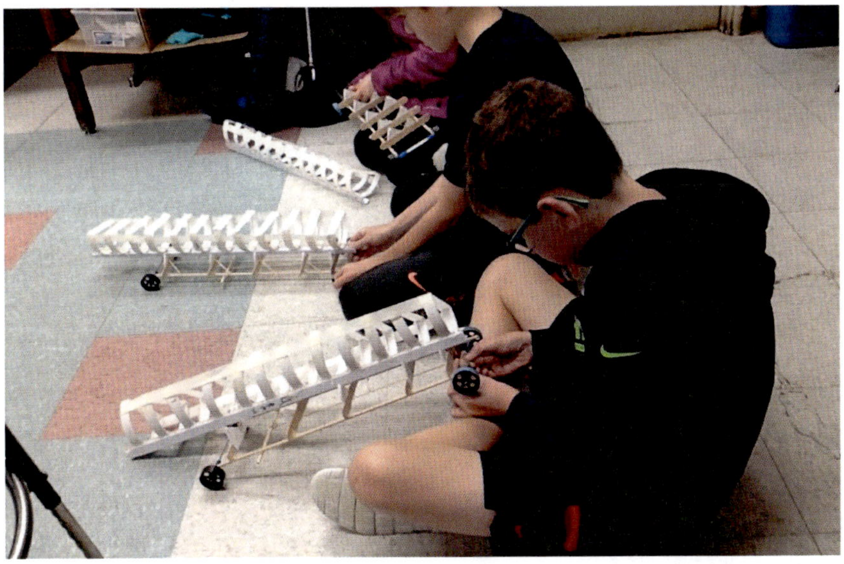

Count the tiles.

How many
was that?

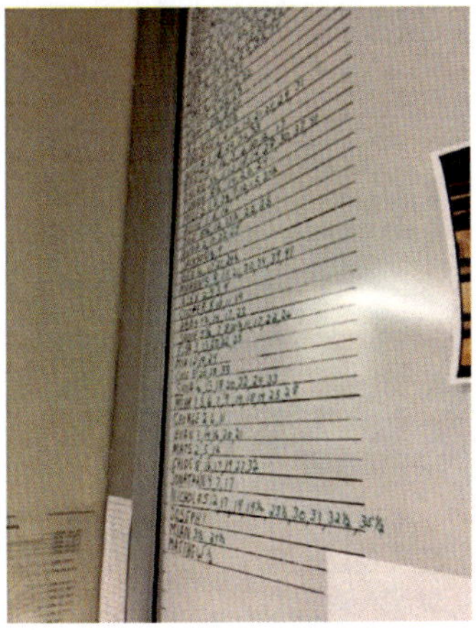

Suddenly names go up on the board.
My rig moved two floor-tiles.
Mine moved seven.
Witness, witness!
Can you verify that mine moved 21 tiles?
"Clear the course!" "Witness, witness."
The room fills with excitement.

People began to change wheel size to try to go farther and farther.

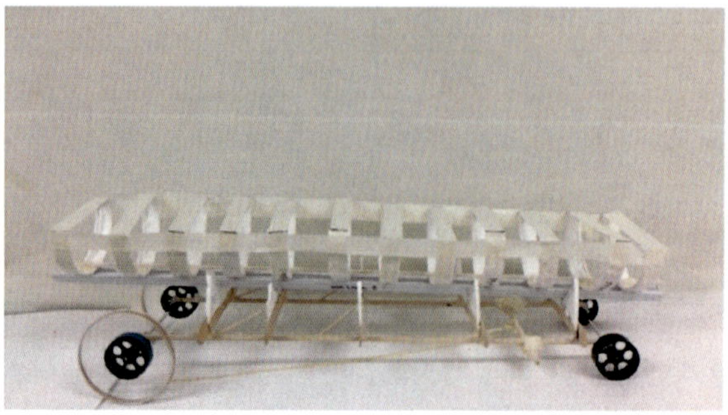

Petri dish wheels

And soup lid wheels

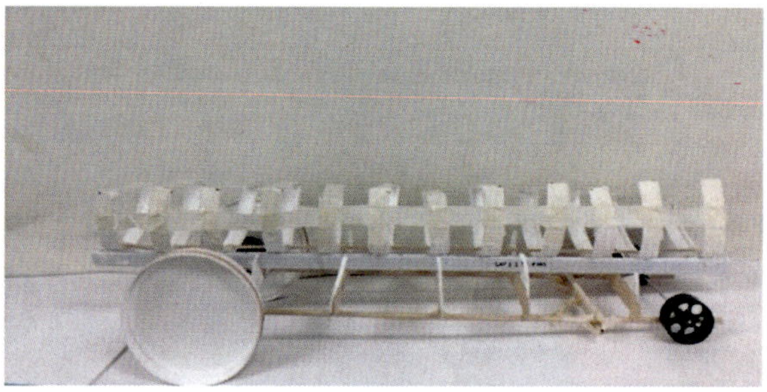

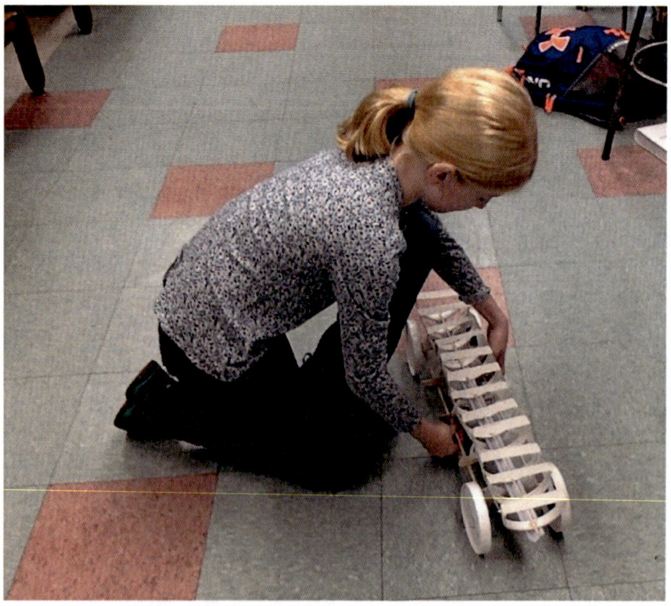

Doc "O" says it is time to build a cardboard hull.

Remember, we will make our ship
about 2' long, 4.5" wide, and about 2.5" to 3" deep.
ULine sells boxes to store small parts in plumbing shops.
Item S-2718 is 4"x24"x4.5". That is just about the right size.

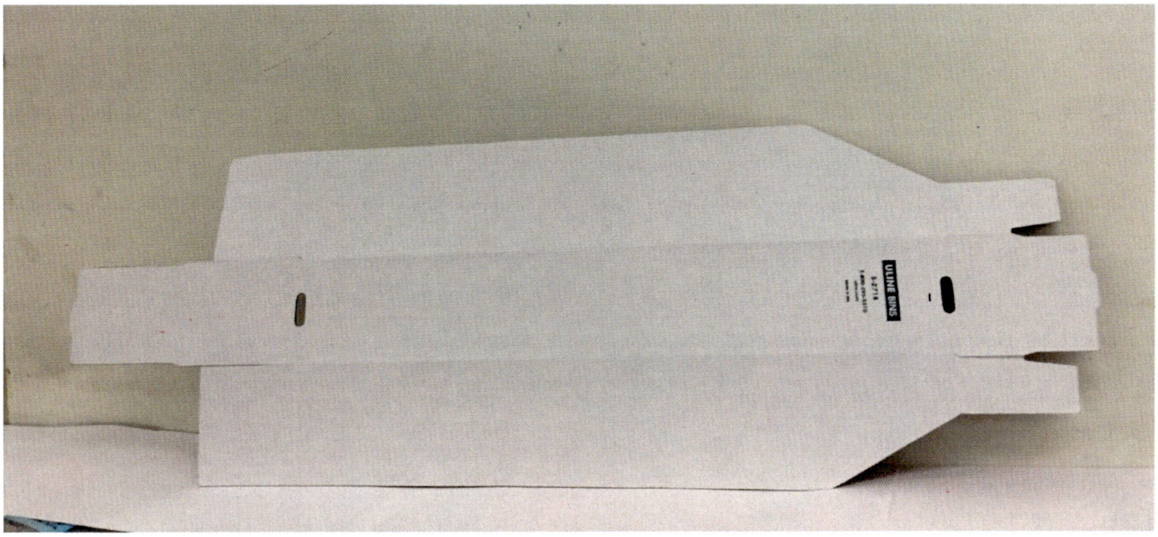

We can turn it upside-down to build our ship.
We can make a paper pattern to cut out the bow.

First we have to learn to fringe and fold paper to shape the hull.

The paper pattern worked pretty well so we are ready to try to shape a cardboard box.

Bow
Stern
Tumblehome.

What's a tumblehome?
Well, ships had round-sided hulls. The roundness was called tumblehome. The widest part was down at the water level. That way shrouds and ropes could attach to the sides of the ship without rubbing on the gunnels. That also kept the decks of fighting ships apart so men could not jump from one ship to the other.

Boxes are hard to bend but we can do it. Students have strong hands. Some people are really good at making round hulls. Tumblehome.

Bend up the side.

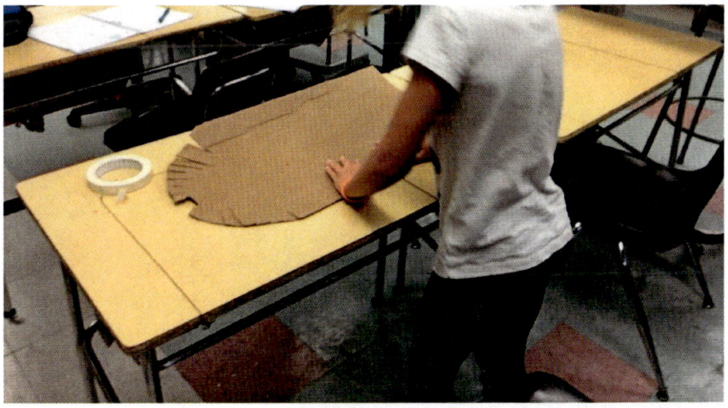

Now grab the other side.

Push down
against the table
and bend the box.

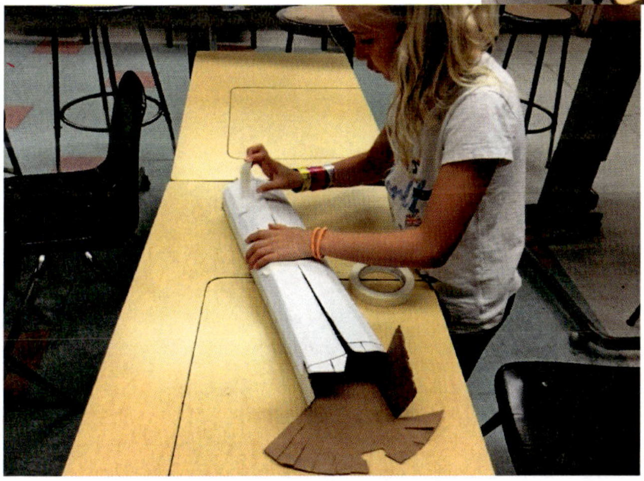

When both sides are rounded
bring the edges together
to form a keel.

Tape it in several places
before you glob glue on it.

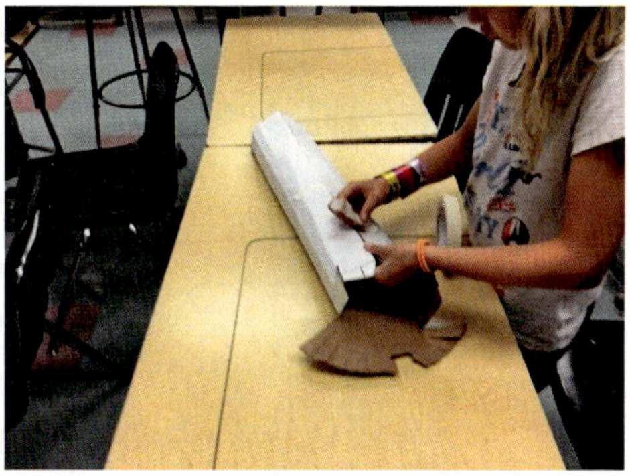

The stern has to be cut
in curves so the transom
will fit.

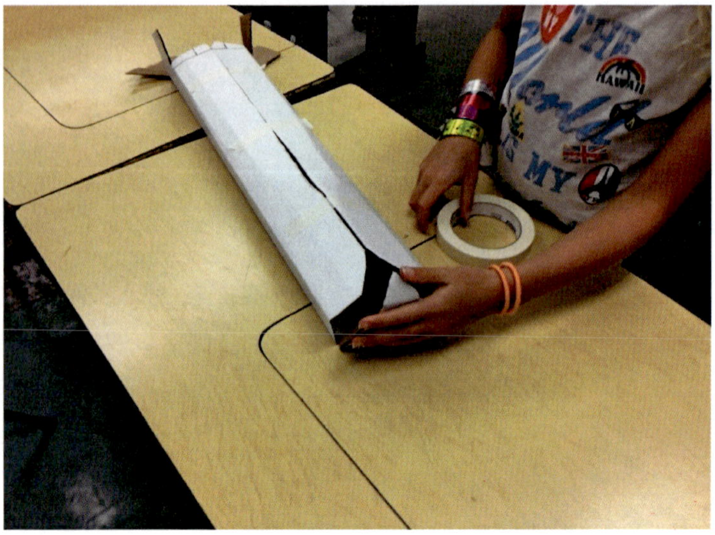

Fold.
Tape.
Get ready to glue.

Everyone figures out
how to fit the pattern
and cut the box.

Cut and bend.

So now to glue the hull together. Make it round.
Tumblehome.

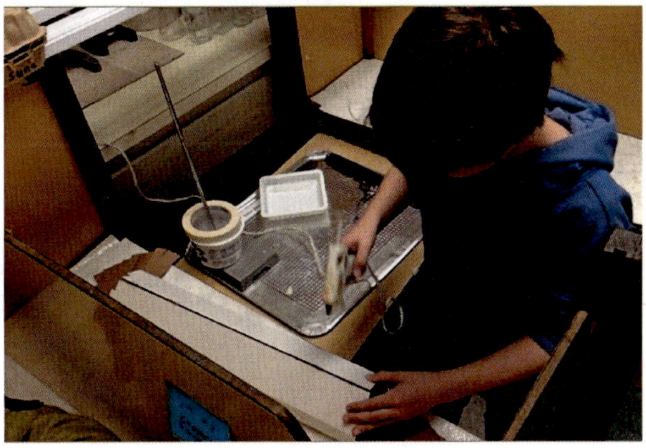

Lots of glue.

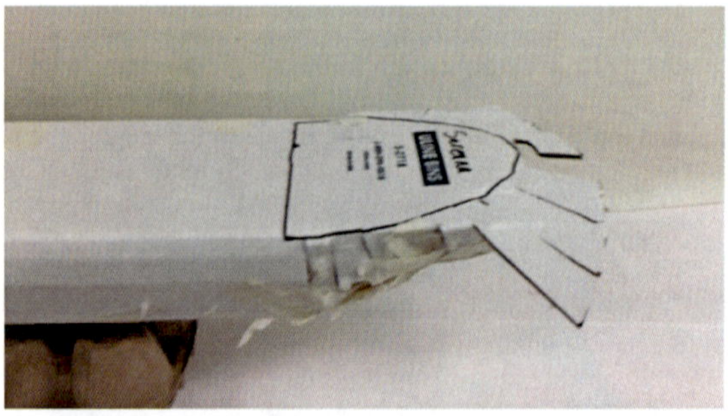

Then we have to fold in the fringe sections to make a bow.

Use masking tape to stick the flanges together.

Tape, tape.
Watch the bow taking shape.

Take time.

What a beautiful, neat bow.

Now to add some glue.

More glue.

It is starting
to look like a bow.

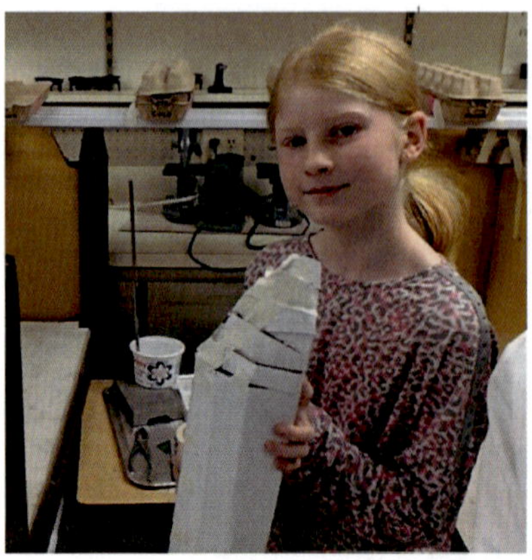

Oh yes! Pretty cool ship hulls.

But the bow needs more tape. And glue.
Like Paper Mache. Or a mummy.

We have to put on three masts. We know masts are "stepped" by running them down through the deck so the foot of the mast sits on the keelson. Our masts will be three-eighth inch dowels. We want to be able to take them out – at least until we glue them down with the shrouds. We need something to hold them. Oh! Drinking straws fit around the dowels. We can glue cut-off drinking straws in the hull and slide the masts into them. They are located 1/5,1/2 and 3/4 of the way from the bow.

Fractions are tricky.

Our ship is 24"
So
1/5 is about 5"
1/ 2 is 12"
3/ 4 is 18"

all measured
from the bow

Measure carefully

Long metersticks work better than rulers.

Some people put in drinking straws first so masts could be taken in and out.

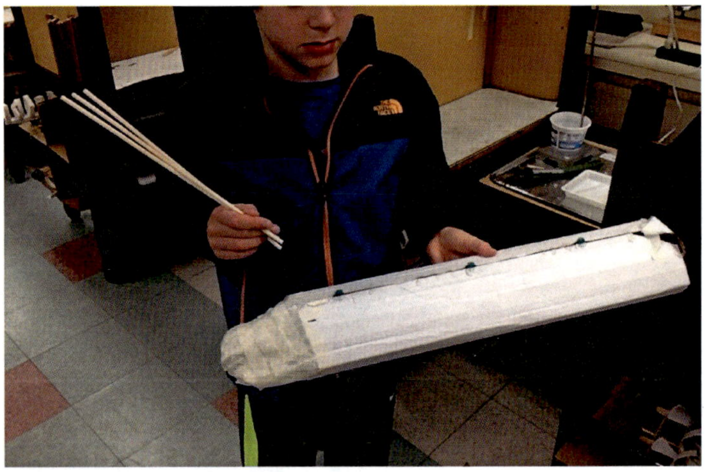

Some just rammed their masts in place.

Everyone loved the idea of stepping masts.

After the masts, we put in the bowsprit.

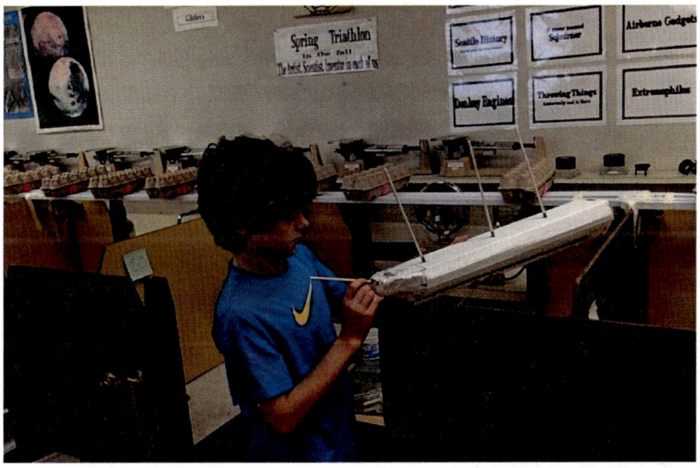

The bowsprit
should stick out 6"
It is a 12" dowel
so
6" is slathered with glue
and inserted in the bow.

Some people
took the time
to peal another j-strip
to make a keel
for the box hull.

We need to tape the
ship to egg cartons
so they stand up
while we work on
them.

Egg carton blocks
and shores.
With masking tape
seat-belts
to hold them on.

An even fancier bow. This one has two seats-of-ease in the beakhead on the bow.

Sailors had to use these outhouses without walls. The splashing waves cleaned their toilets. Putting them on the bow put them downwind from the ship and blew the smell away. Pretty clever.

The bowsprit shrouds and bobstay go on. A bottle cap drilled with a drill-press makes a nice cap. The cap has two holes so the jibboom can slide in one and move down along the bow sprit.

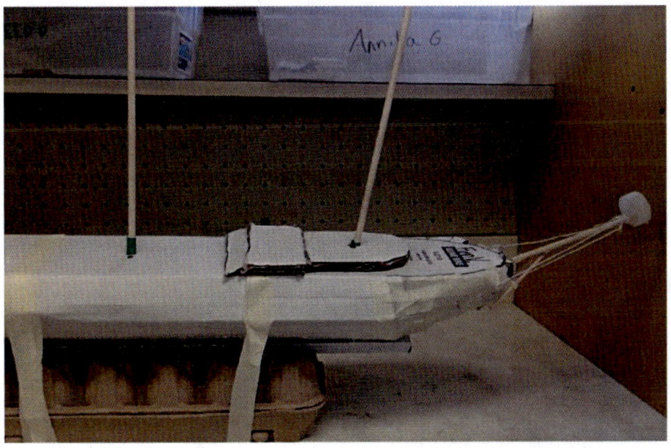

This rig has a pretty accurate Fo'c's'le.

Our drill press has a coke bottle on it so our hands don't slip in.

Doc "O" keeps the key in her pocket.

After we put in the jibboom, we glued on the dolphin striker and martengale.

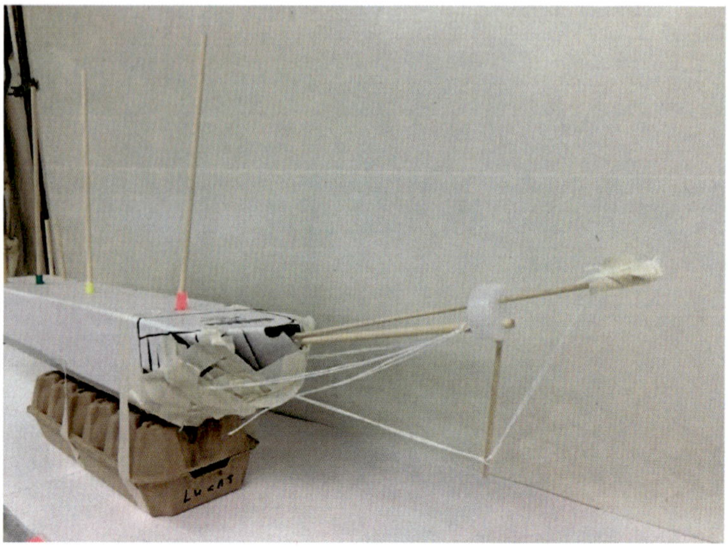

Horses sometimes wear martengales to keep their head down so they can't buck. Sailors for these ships often came from farms where they worked with horses. I guess the bucking seas reminded them of bucking horses. The ropes used to strengthen the bowsprit and keep its "head down" in a storm was just like a martengale back at the farm. It was an obvious name to use.

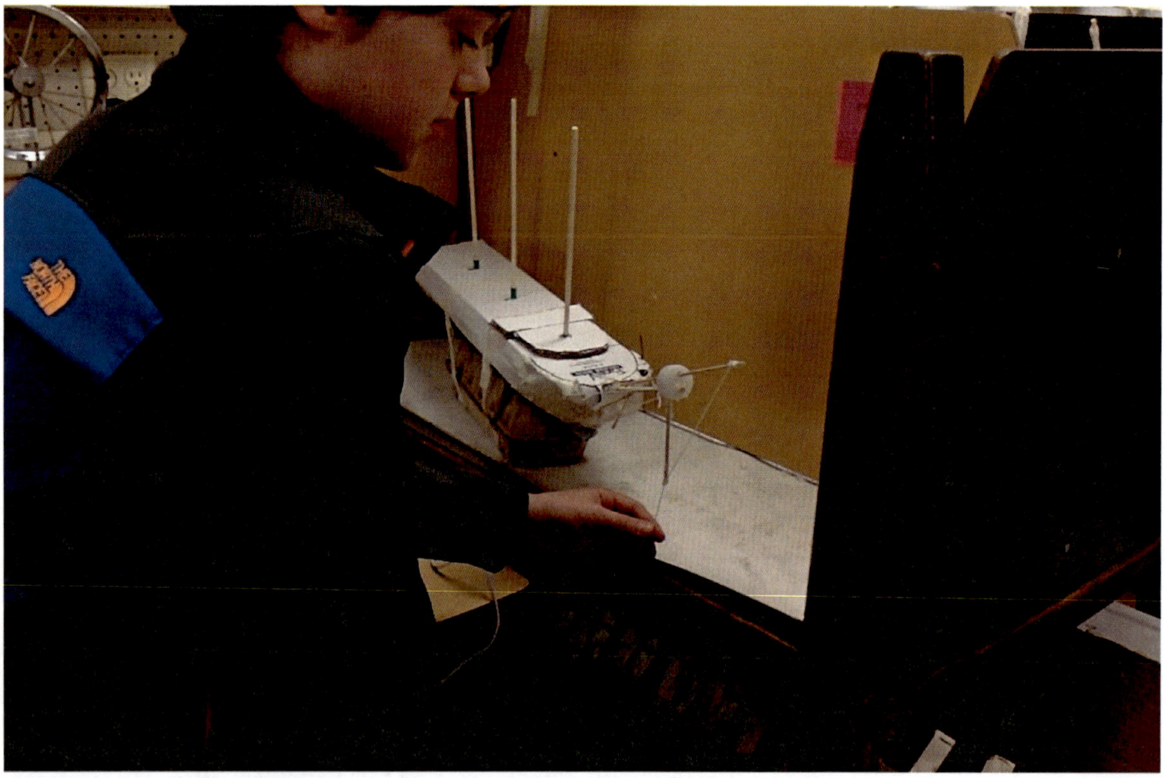

We think the dolphin striker spear point is a joke.

The purpose of the dolphin striker is really to make a truss out of the bowsprit / jibboom assembly to make it more rigid. The martengale helps with that.

Before work on the shrouds many students wanted to build some more decks and cabins.

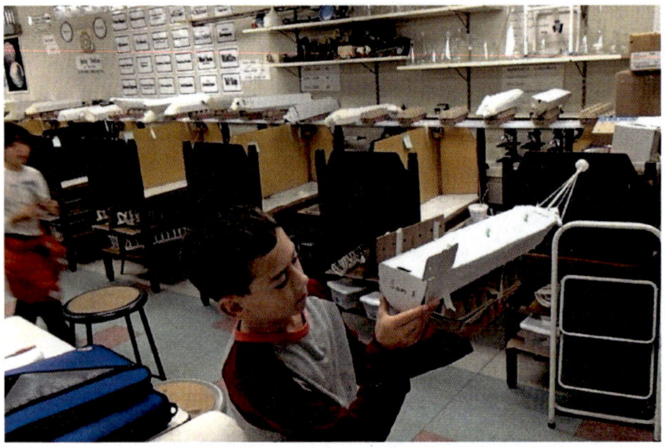

A poop deck goes on.

And another.

The forecastle is called the fo'c's'le, The stern deck is called the poop deck. What an odd name. Some people make a pretty poop deck several stories high.

Both a Fo'c's'le
and a poop deck.
Along with a jibboom,
dolphin striker
and martengale.
It is looking good.

Some poop decks are really big.

But now we need shrouds. Four on the foremast, five on the main mast and four on the mizzen mast. Before we can put on shrouds we need to build the "Top". The trestle-trees and cross-trees support the top. How can we hold the tiny bamboo skewers in the correct spacing to make them? Aha! Berry baskets. The very thing! Berry baskets have their webbing spaced just about right for trestle trees and cross trees. Glue the skewers to the berry baskets and then cut away the basket. That leaves a nicely structured top. First though, we need to add cheeks to support it all. We will use little cardboard triangles for cheeks.

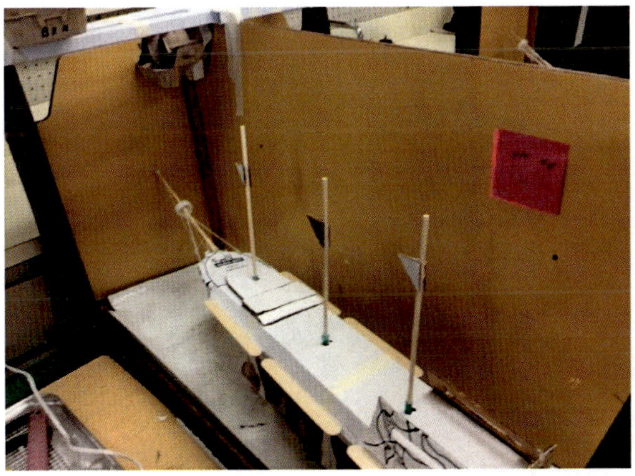

You have to hold the "cheek" triangles while the glue hardens.

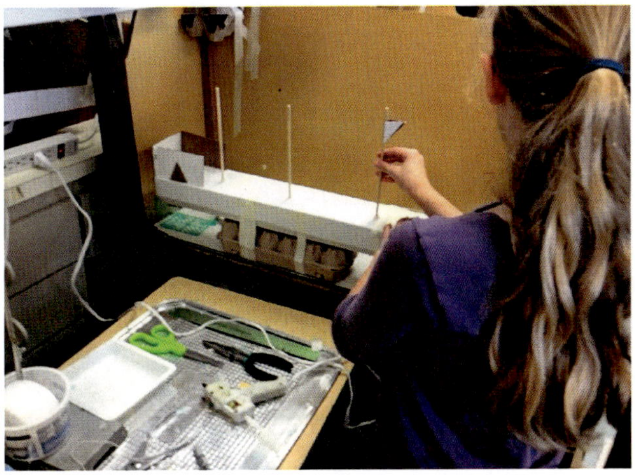

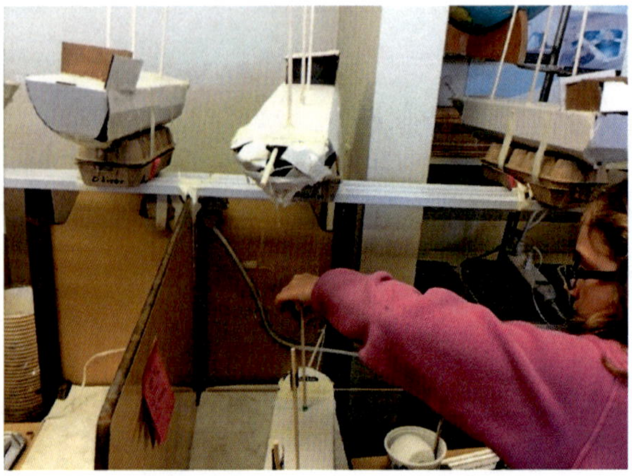

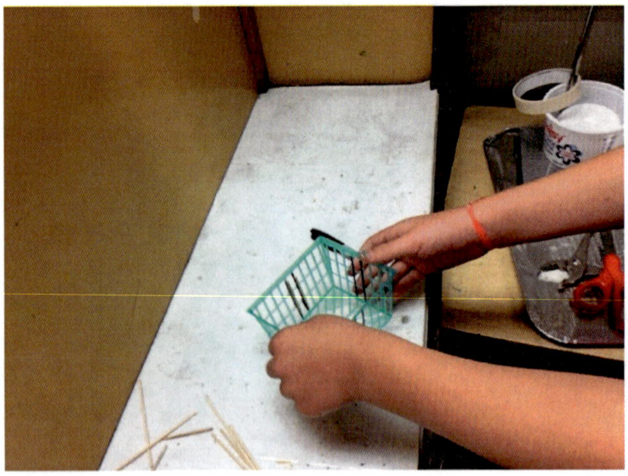

Now for the platforms
That sit on the cheeks.

The platforms are called
"Tops". They are made of
trestle-trees and cross-trees.

Short skewers are the trestle-trees.
They go on the middle webs of the
inside of the berry basket.

Daisy

Next the cross-trees
are glued horizontally
to the long webs
on the outside of the basket.

Let the hot glue harden.

Then cut out the basket side.

Tall Ships

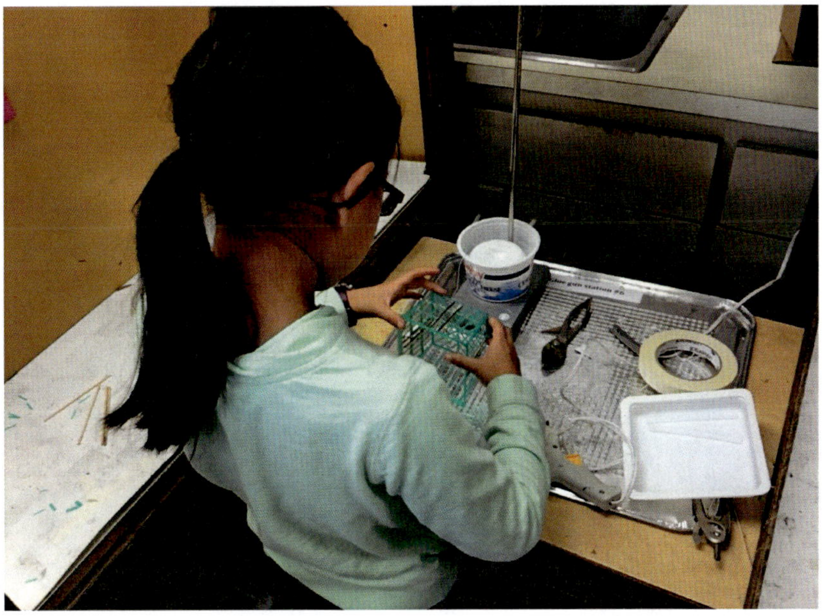

When the trestle-trees and cross-trees are cut out, set them over the mast and glue them to the cardboard cheeks.

Add glue
to make sure
they stay level.

Pay attention!

Make sure each
one is level.

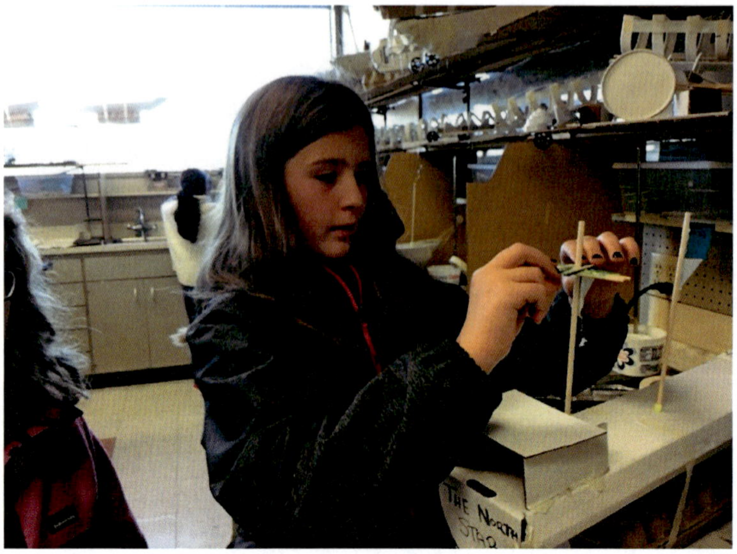

Just about perfect.

Glue it level.

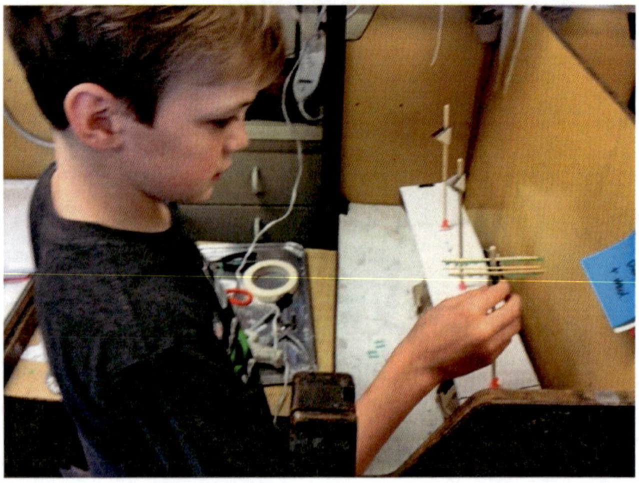

Be sure the long skewers go across the ship.

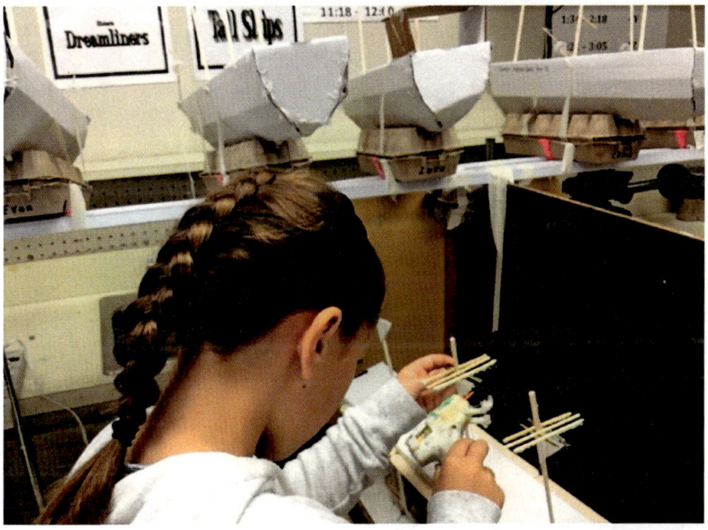

Careful work!
Basket by basket,
we are
making progress.

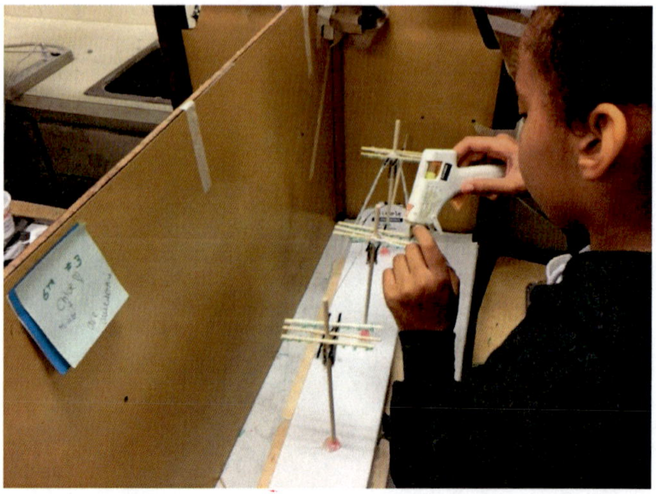

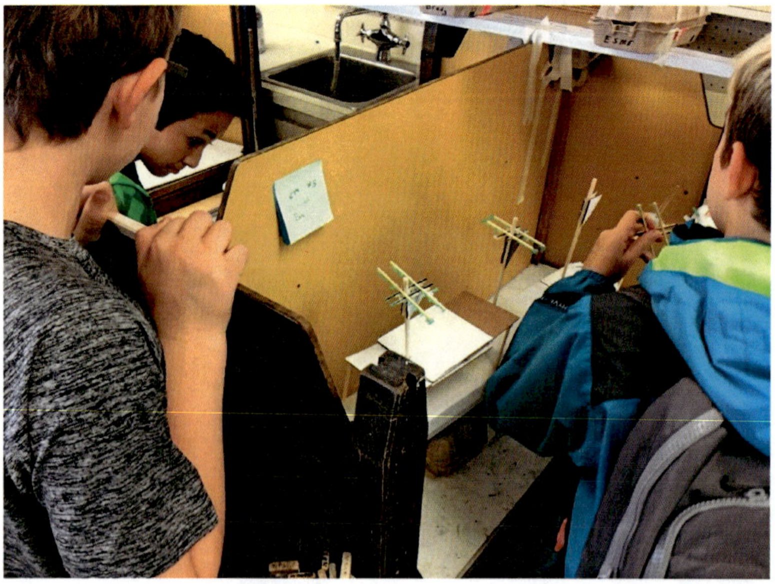

Shrouds start on the mast just above the top but where do they end? Chain wales. Channels. We need to add channels along the hull for the shrouds to attach to.

But what should we use? Tongue depressors work just fine.

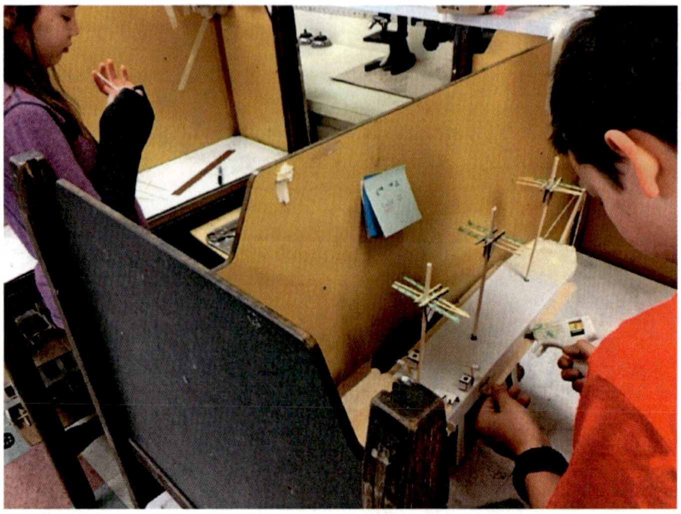

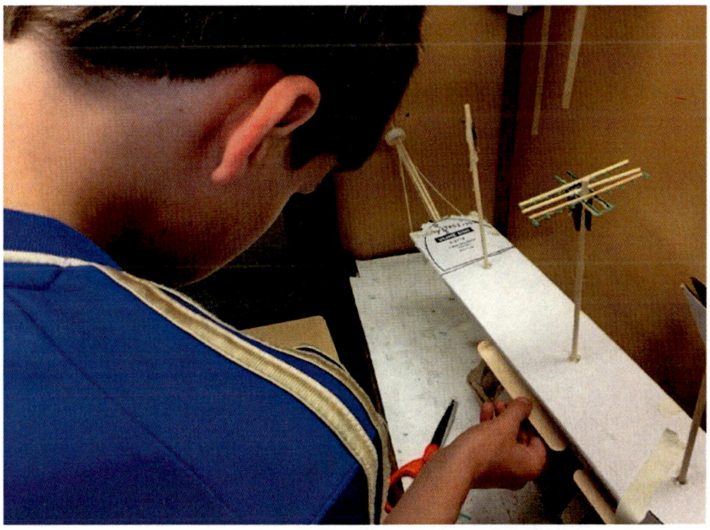

You have to be careful to get the channels placed just right.

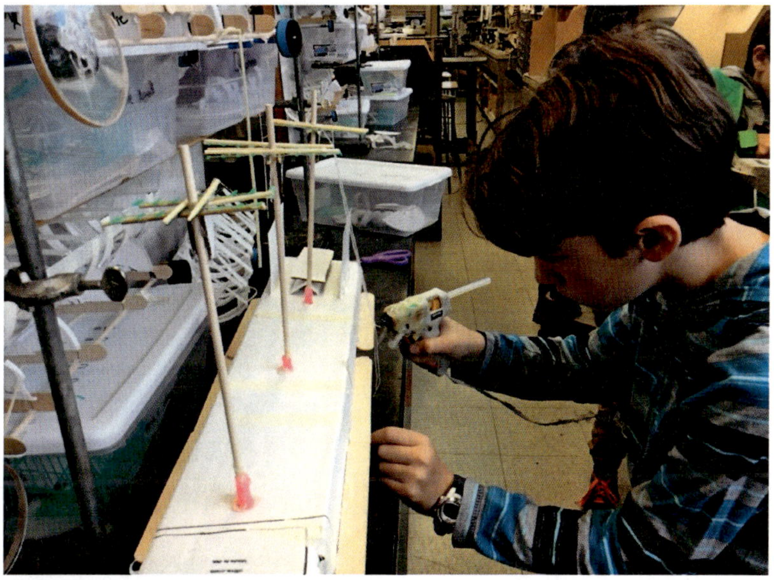

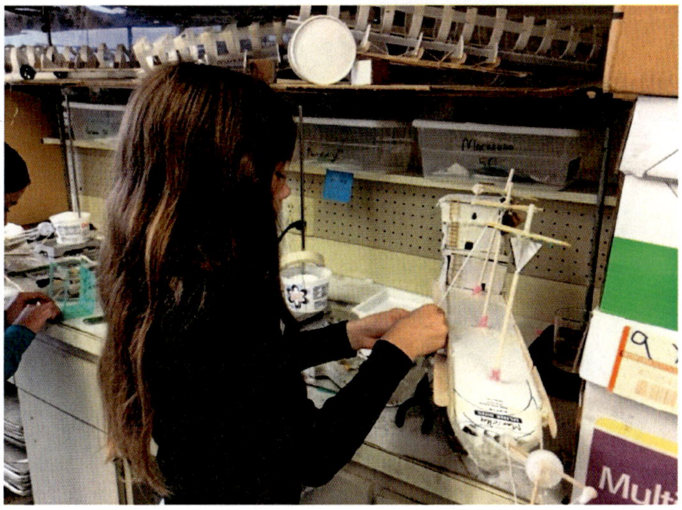

Once the channels and tops are on we can begin adding the shrouds.

It is a bit tricky to learn to tie an overhand knot.

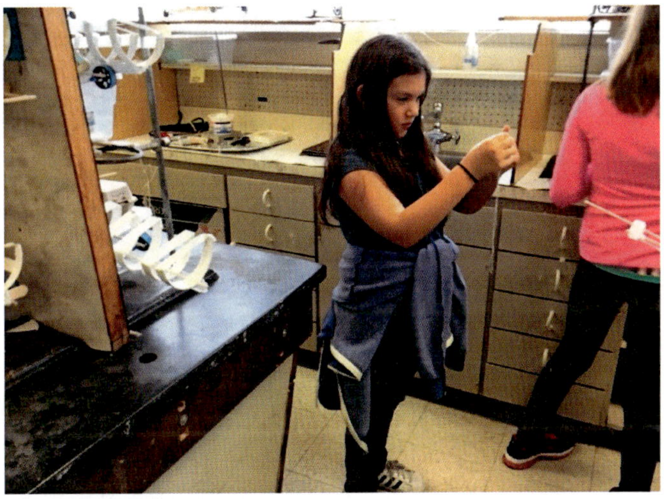

When you learn to tie,
you become a "Local Expert".

Then you can help others.

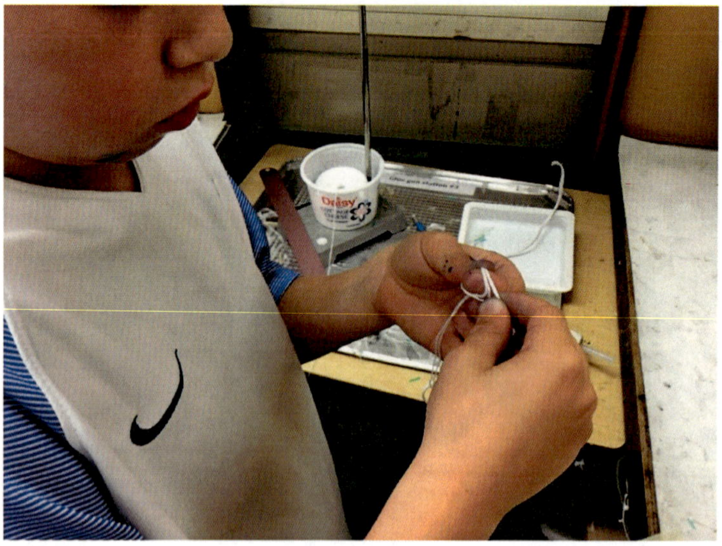

Lets see. Around the finger
then the end goes
through the hole.

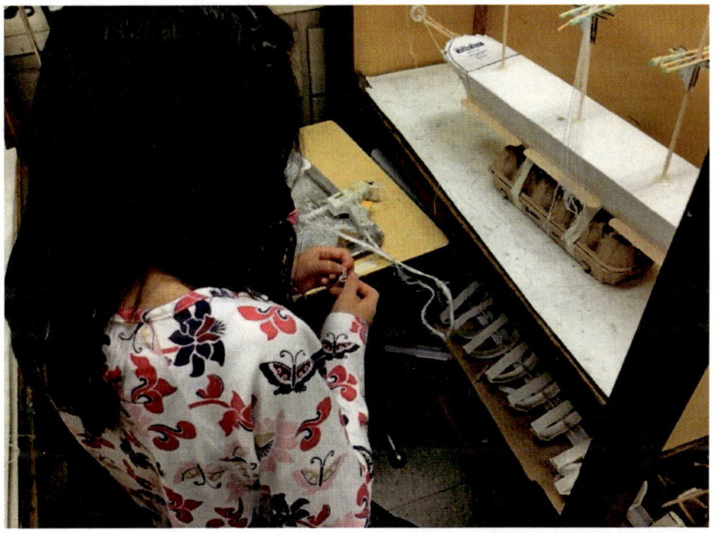

Slow and sure.

And patient.

You have to have
a long enough string
to make each shroud.

The knot is tied
and the shroud goes
around the top of the
mast.

Then pull it down
between cross-trees.
and glue the end
to the channels.

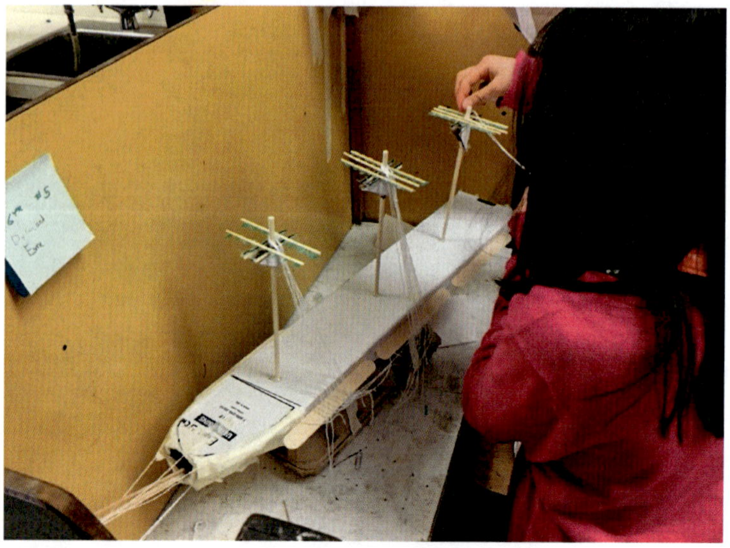

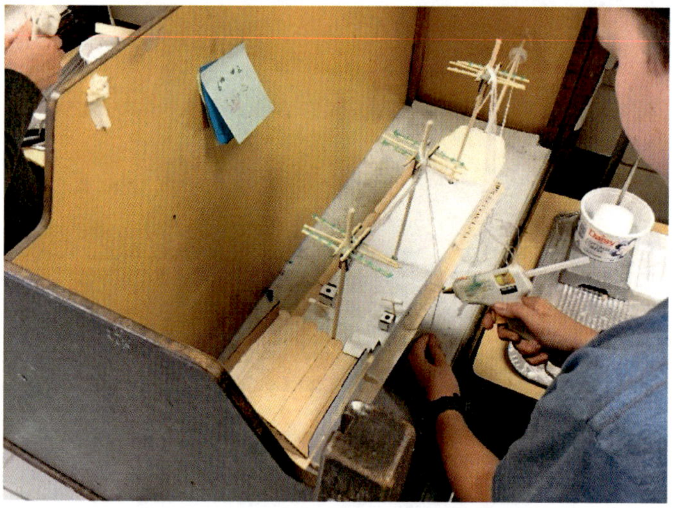

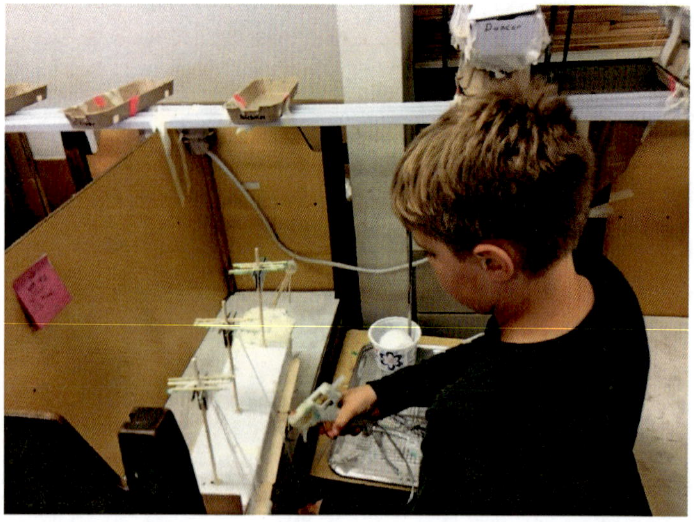

4, 5, 4
You have to make
13 shrouds per side.

That makes 26 shrouds in all.

And that is just for the lower masts.

Make it tight but not too tight.

You need to get the tension even from one shroud to the next.

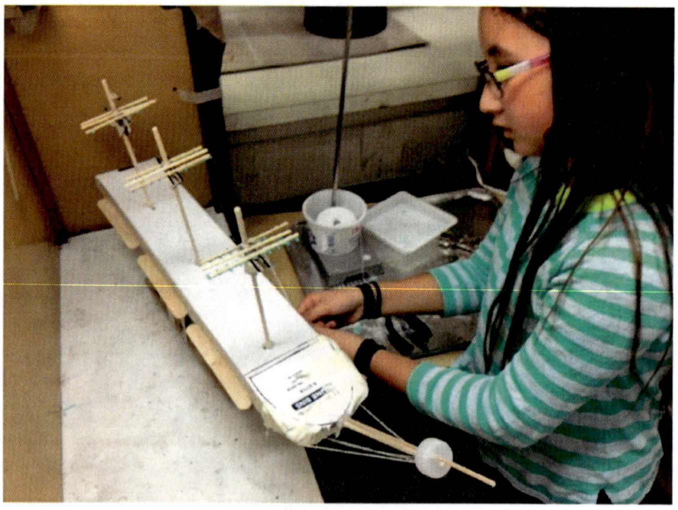

Stays go front to back.
The forestay ties
Just behind the
Mast in front of it.

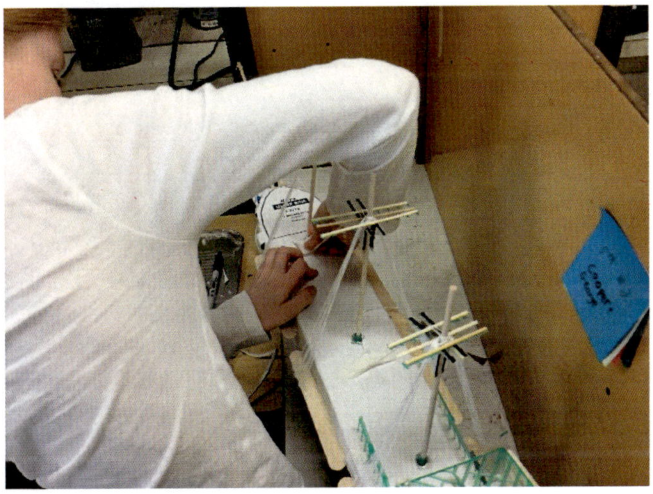

Measure the needed length.
About two lengths of a ruler.

Let the shroud ends dangle. We have to attach them back to the hull below the channel.

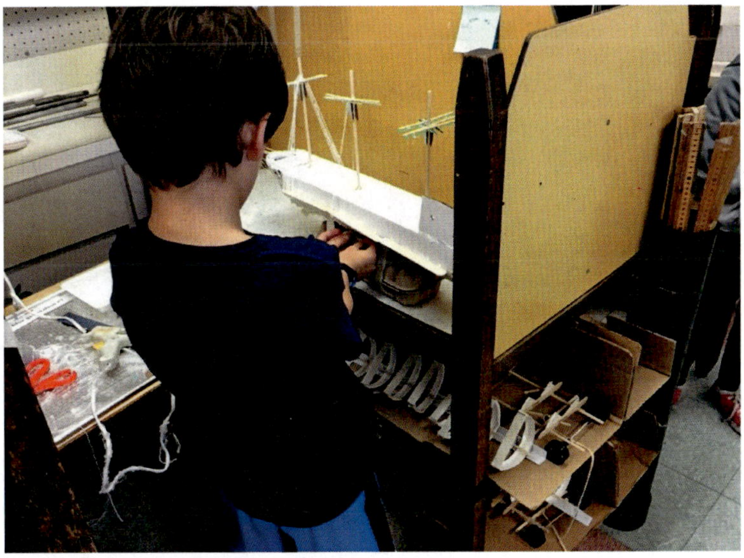

Shrouds have to be neat.

The two ends need to be separated before they are glued to the channels.

Lots of string.

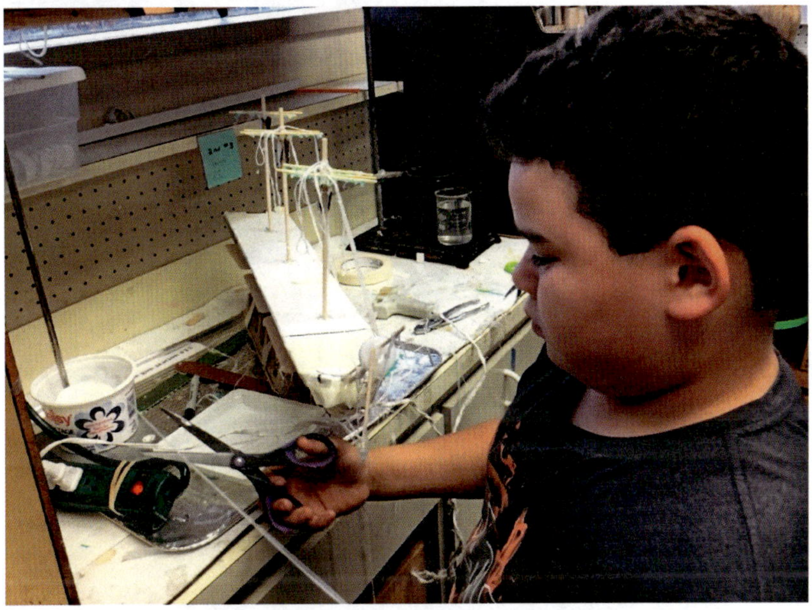

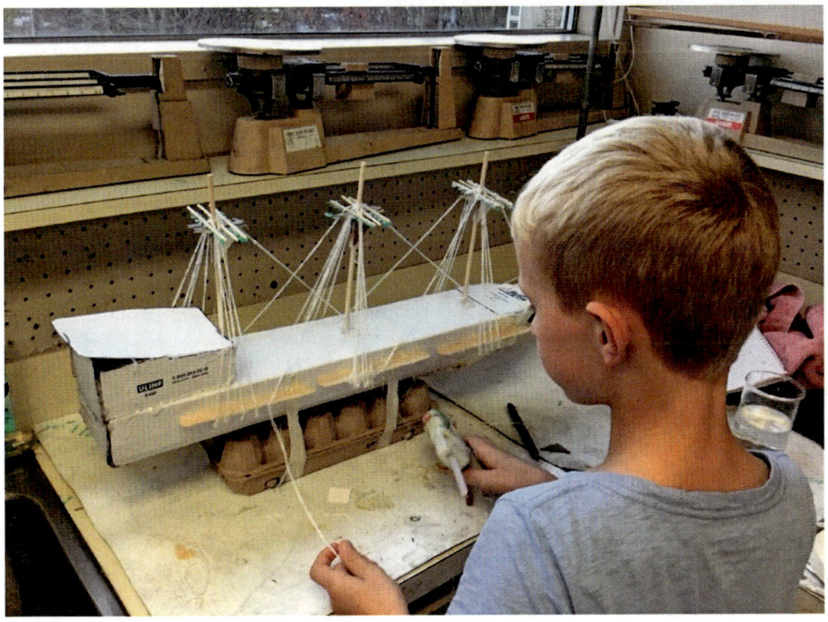

Shrouds start beside the mast and then line up behind that toward the stern.

That is because you will need space for sails in front of them.

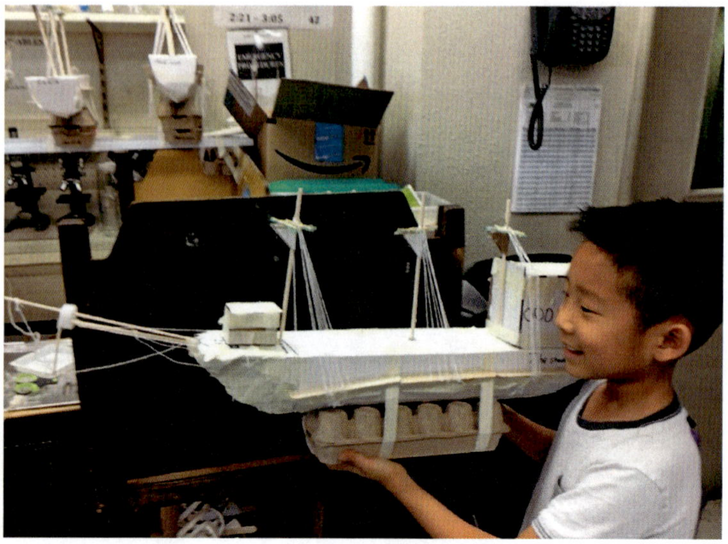

How can we keep the shrouds from sagging?

It helps to tip
the ship when
we glue the shroud ends
to the hull.

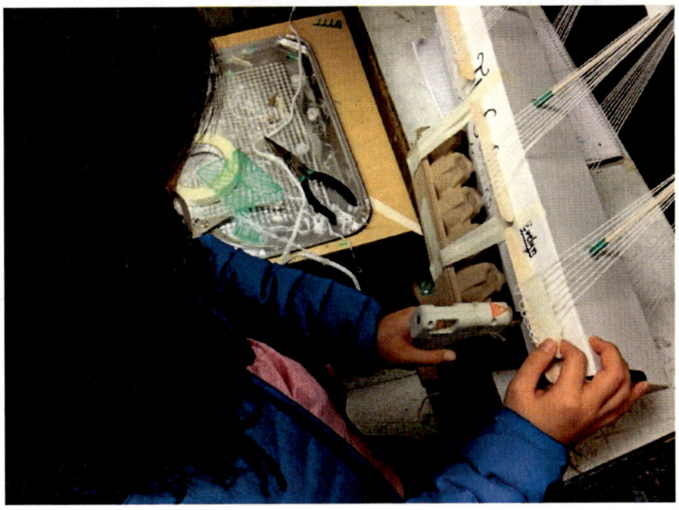

And trim off the ends after
shrouds are glued to the hull.

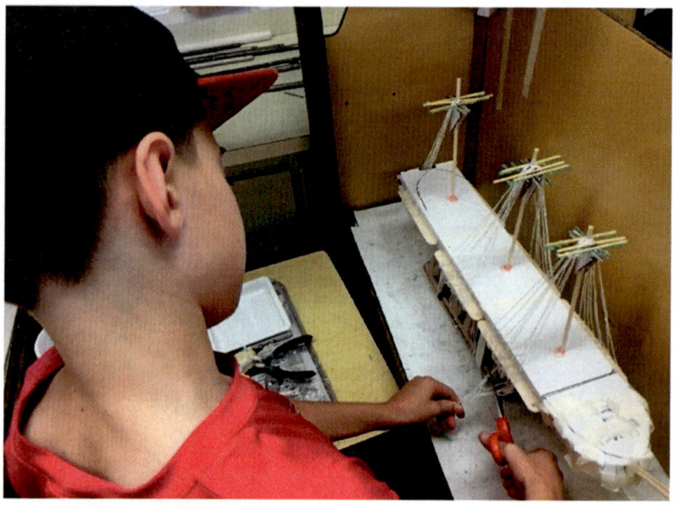

Be careful with every knot.

Balance the ship.
One on one side then
one on the other.

Tinkering Tall Ships

"Local Experts" are willing to give advice.

We take that advice and shrouds go up pretty well.

Position and spacing matter.

Tight shrouds look good.

Nicely spaced.

Now glue them to the hull.

Work, work!

Making Progress!

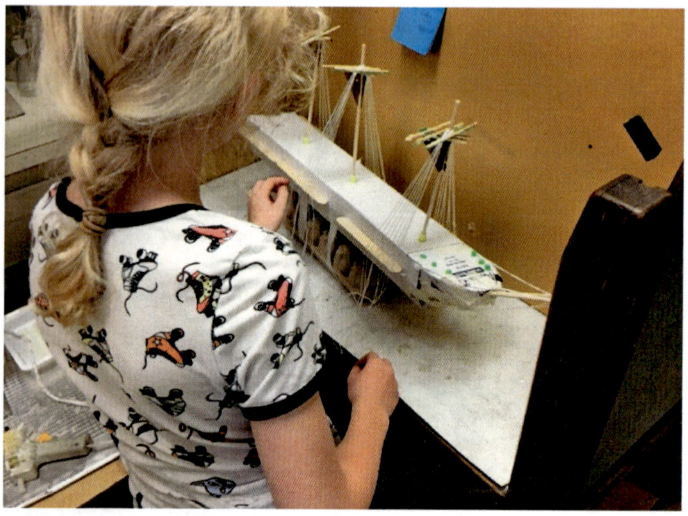

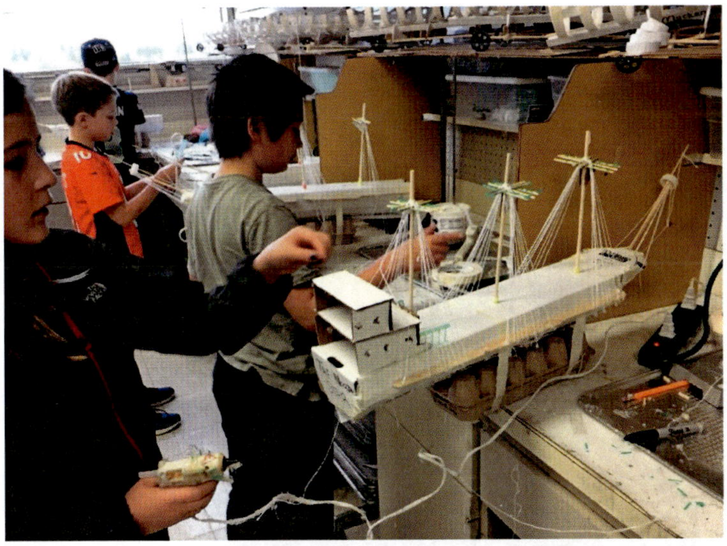

Better add finishing touches to cabins before shrouds make them hard to reach.

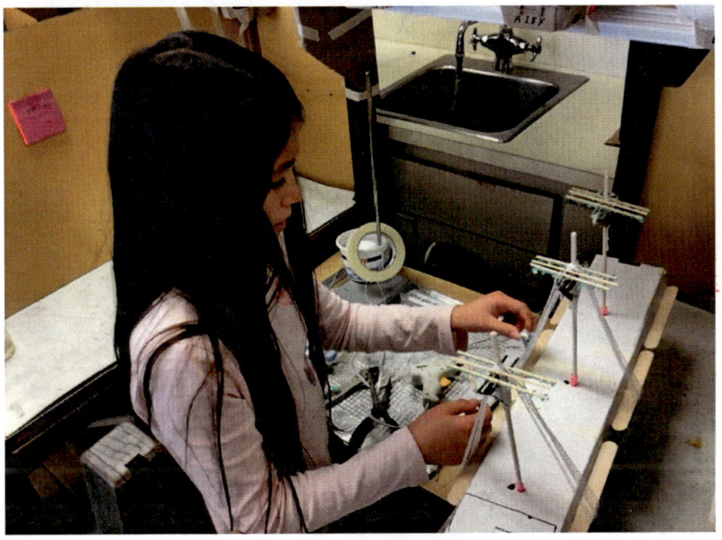

Gluing shrouds takes focus.

Tie the knot.

Pull down carefully.

Squirt glue and
wait for it to harden.

26 shrouds in all on the
lower masts.

Three more caps are drilled.
Two holes in each cap.

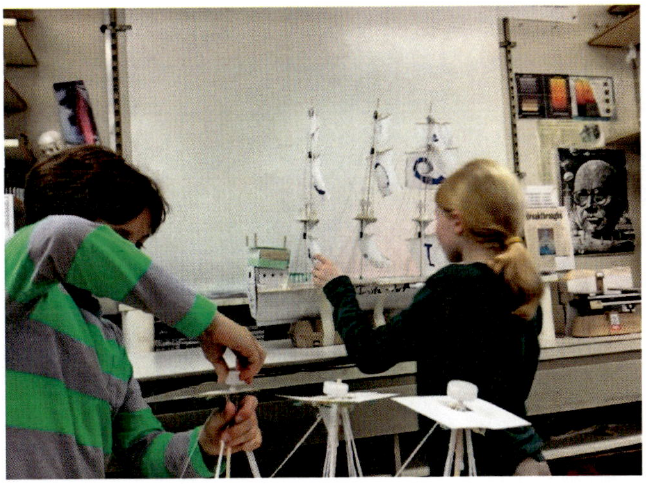

They are set on the very top of the mast.

Top masts go through the second hole in the cap. The masts overlap making a place called mast "doubling". The doubling holds the top mast steady – at least after we add lots of glue.

There are three shrouds on each side of each top mast.

We use small cardboard triangles for cheeks and a tiny brass washer for the top mast top.

The curved side of the flooring of the "Top" faces the bow.

Upper masts
Go in.

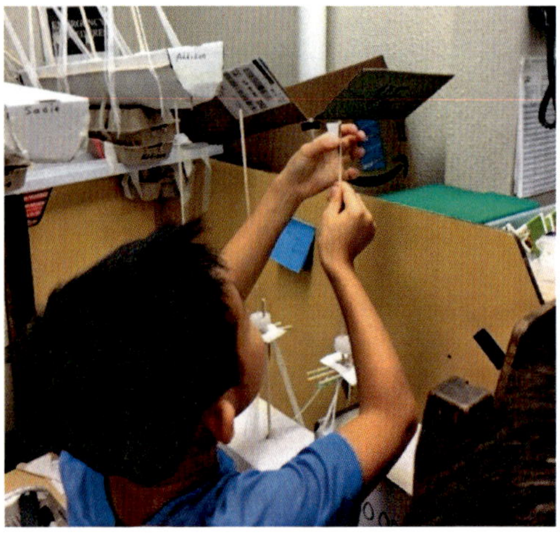

And cheeks

Tiny cheeks
for the top mast.

Working on the upper shroud.

OK. Lower and upper shrouds done
.

What is next?

Yard-arms?

What are yard-arms?

Yard-arms have to hold the sails. They are hoisted up and attached to masts with cranse-irons.

We can use binder clips and bent paper clips to fashion a connection like a cranse-iron.

Glue the binder clip and peg it through with a skewer and yard-arms are up!

Oh my goodness! The very first sails are up!

And then more sails.

Nine sails in all.

But what are they made of?

Giant size coffee filters!

The folds flow down making
realistic looking sails.

Sails certainly
Look pretty.

Our ships
are pretty big.

Wow!
That is something
to be proud of!

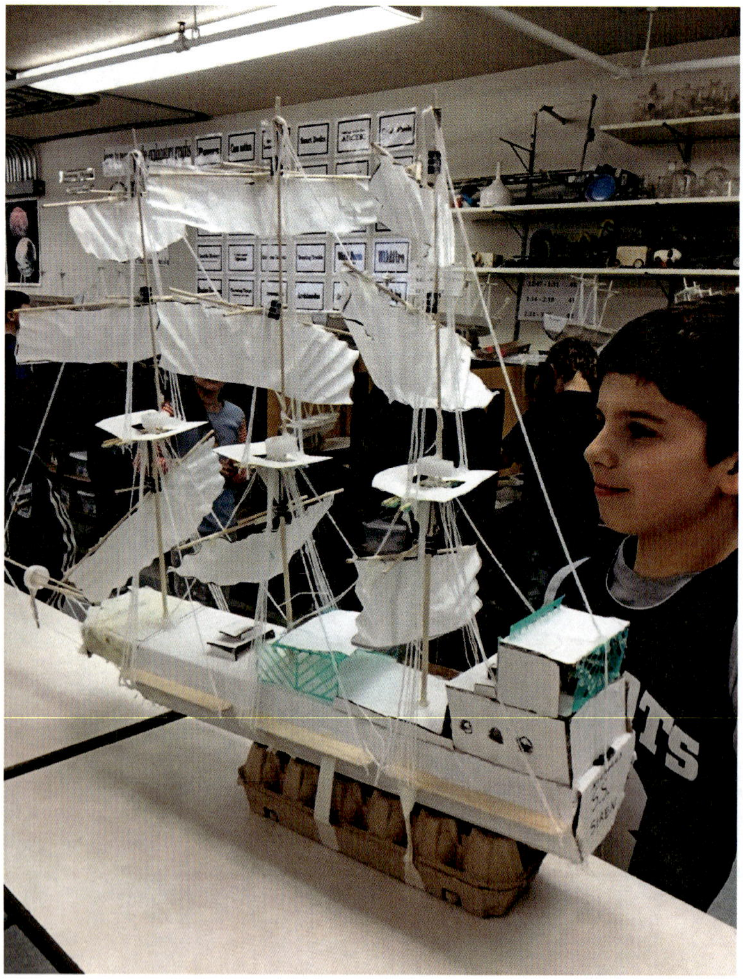

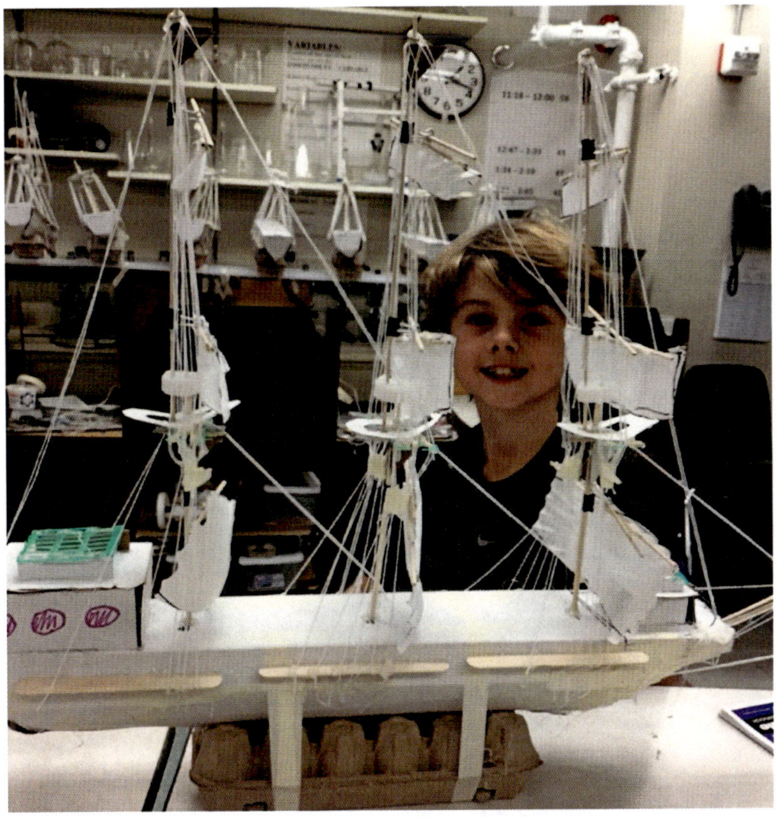

Another one
done!

Hurry, Hurry!
Quickly.

I did it! Before
the end of the period!

We are going to have rig check day soon.
We will have to set our rigs on the cradle and launch them.
Will they move?

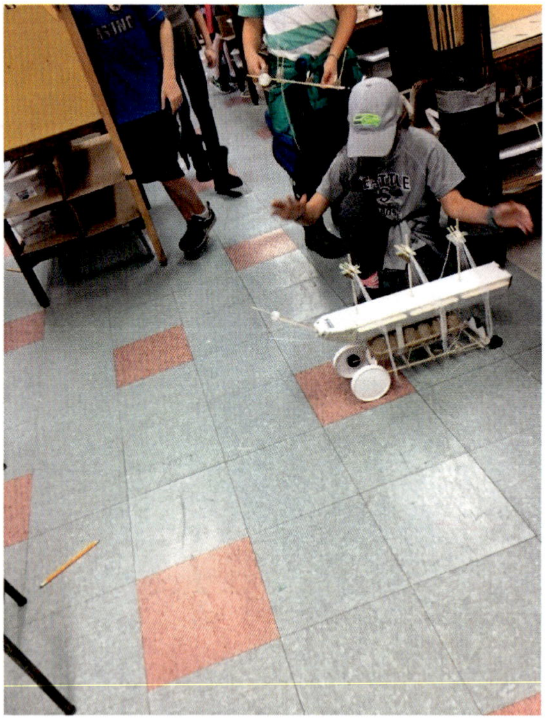

Maybe a new cradle will work better.

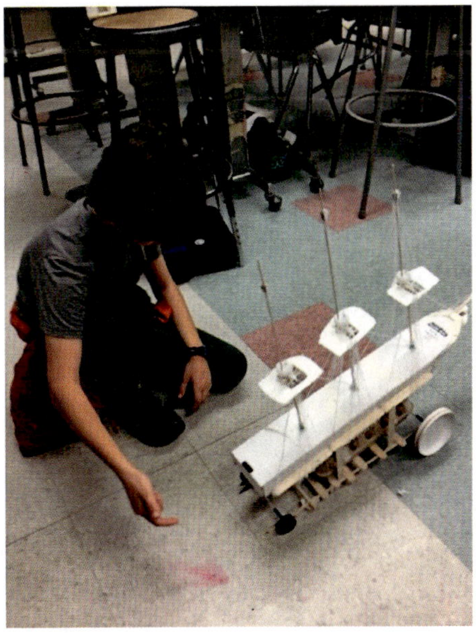

Why isn't it moving??

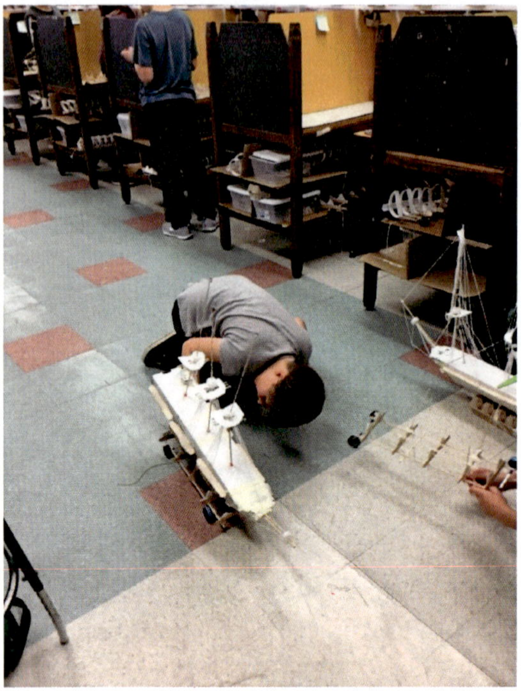

The event day is almost here. Time for practice rig-checks.
Rib rig races first.

Period 1
5th grade is ready.

Ready – set – go.

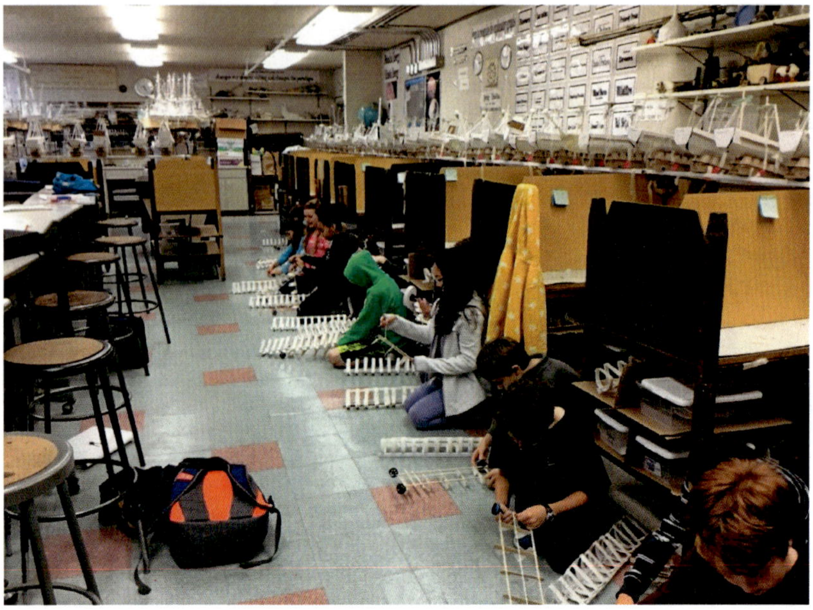

Period 2

Time to practice
for the gym event.

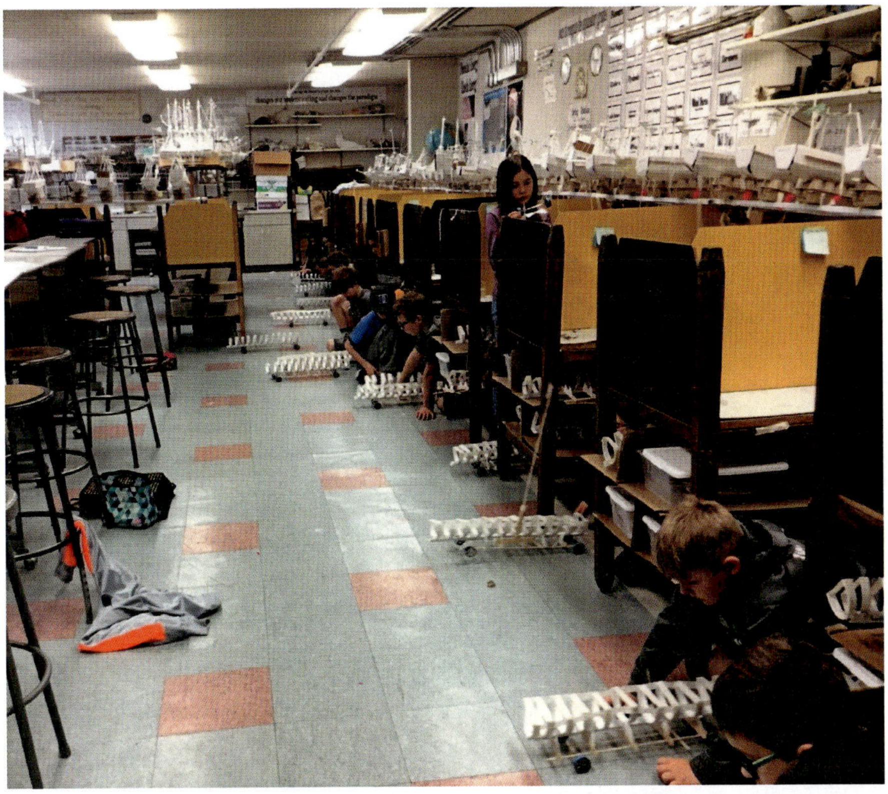

And
Now
For
Period 3.

4th grade
is ready too.

Period 4

Ready, set, go!

Period 5 is next.

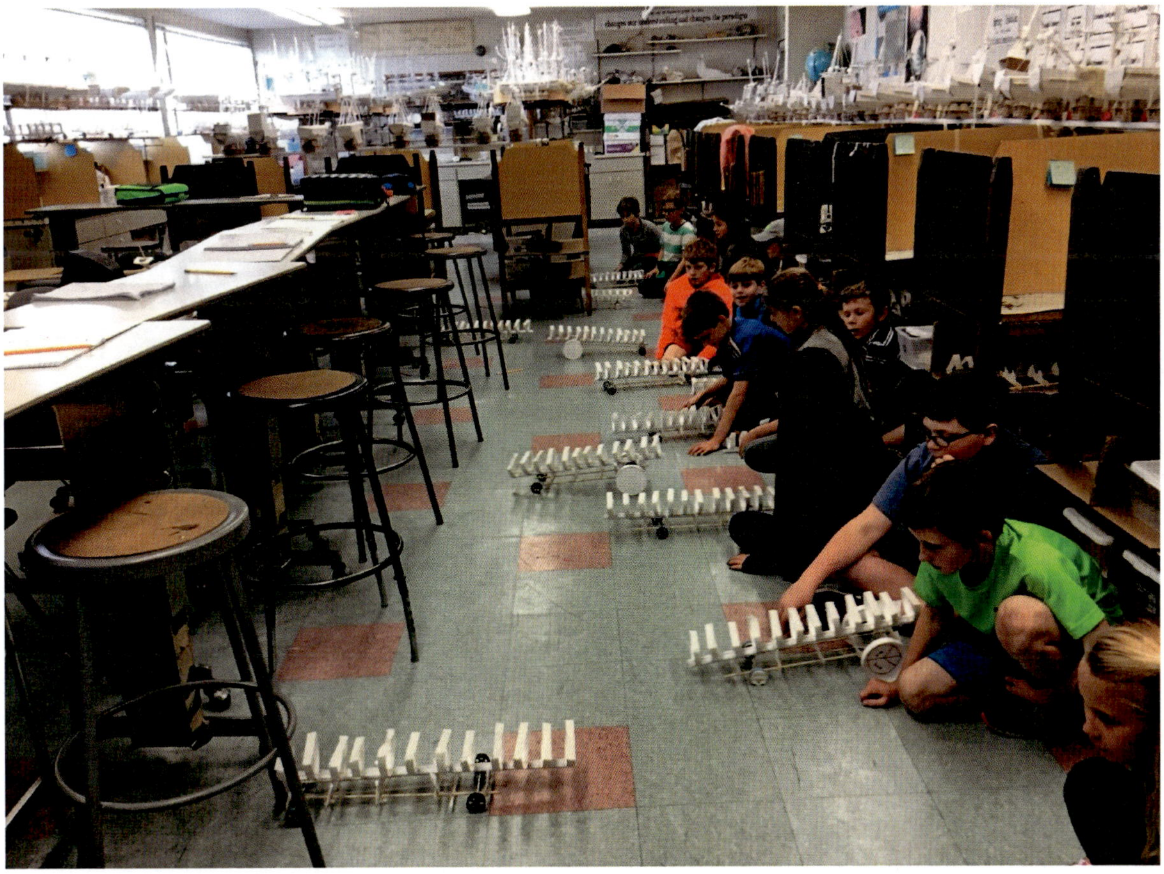

Period 6
practices
Rib-races too.

And next we have to practice launching our tall ships. Rig check day #2.

Period 1

Wind up to launch ships.

Period 2

is looking good.

Period 3

Time for your practice run.

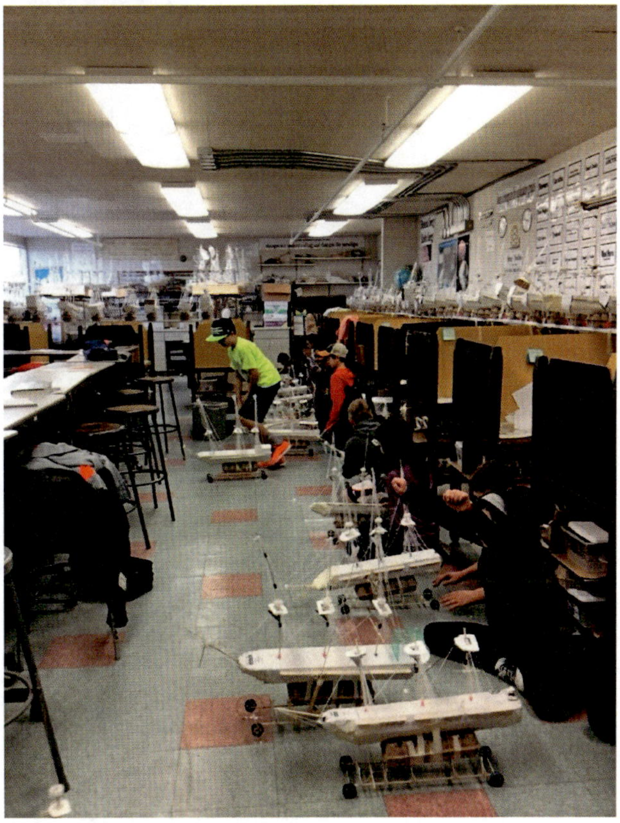

4th grade is a day behind

but they are willing to
practice with their
ship rigs.

Period 5

4th graders
give practice a try.

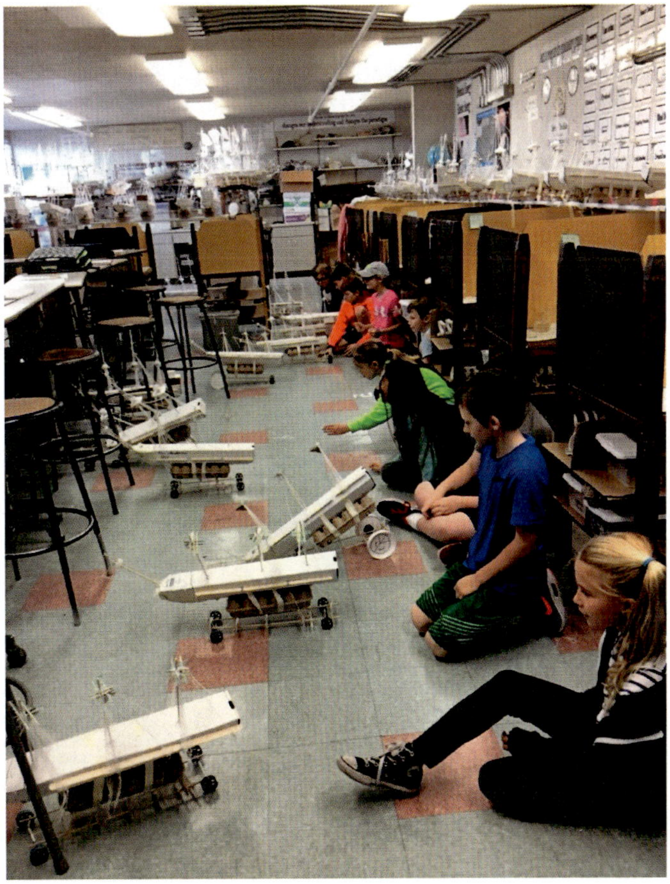

4th grade

Period 6
Practice launching sips.

Now everyone knows
what to expect
when we go to the gym.

New inventions were made the last couple of days before the gym event.
We had an anchor with a working crank and a cannon that shot projectile when you blew in it.

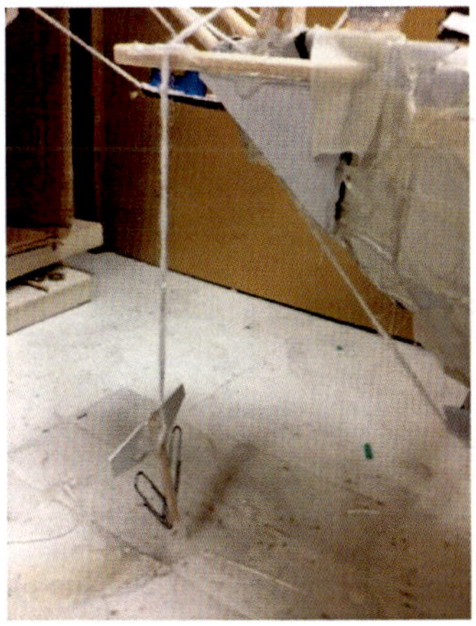

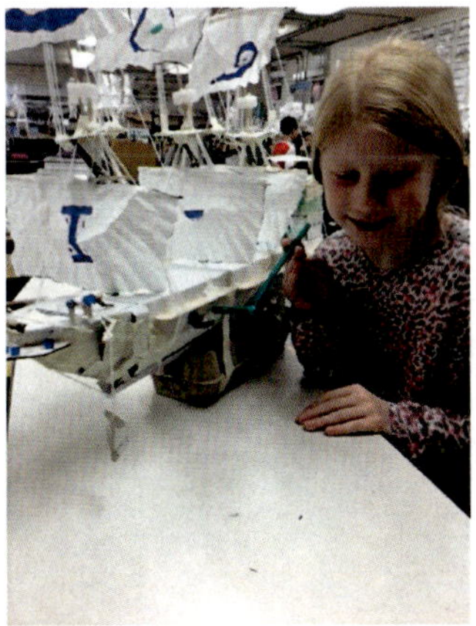

Finishing touches included bringing the top-mast shroud ends in to attach to the lower shrouds. Futtock shrouds. What a strange word. It sounds like swearing. Futtock. Foot hook, Shroud ends that hook around the bottom of the "top".

Yes. There they are.
Carefully done.
Futtocks.

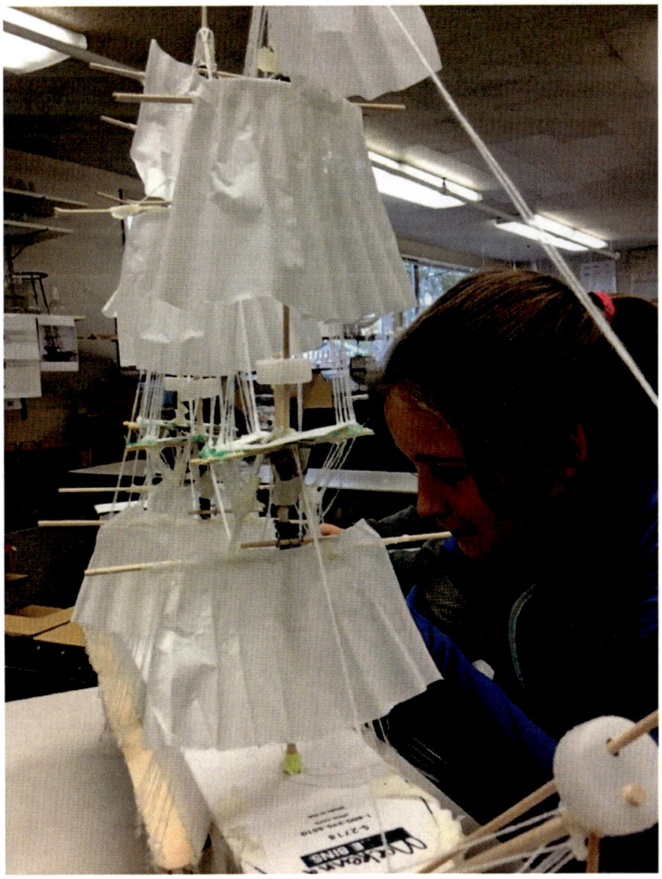

Students began to panic as they tried to finish their rigs before the big event day. Forty or fifty came in at recess to work each day. The room was really crowded. Some added more rigging

Others figured out how to add sails and then helped each other cut and fit sails.

Trace the yard-arm
on the coffee filter.
Outline the sail.
Cut it out and tape
it to the yard arm.

A great example of
a "Local Expert"
being helpful.

Two hundred ships filled the storage racks in the room. Some had lower rigging. Some had upper rigging. Some had cranse-irons and yard-arms. And some had a full set of sails.

Each child is different
by temperament, ability
and maturity.

They go as far as they can go.

Without grades or judgment.

"Finished" is different
for each child.

Each child focused on what
interested them.

Some had more masts and
sails than others.

And that is OK.

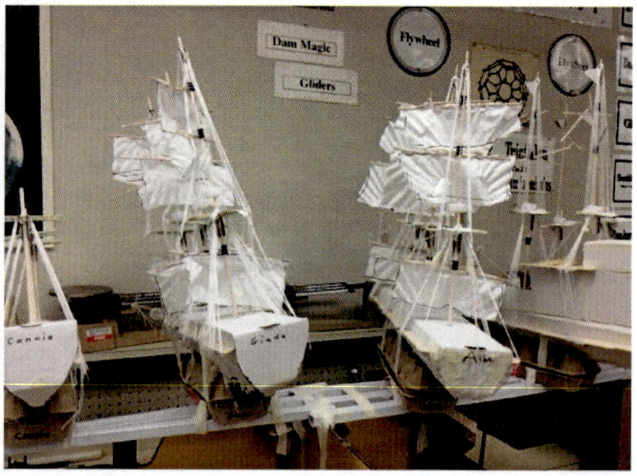
Dam Magic
Gliders
Flywheel

There is no predetermined "finished" state.

It is wonderful .however far they got

And finally it was time for the Engineering Event in the gym.

Posters are stapled to cardboard sheetrock shims and hung on the wall from bent paperclips. Thus – no masking tape on the wall.

The "playing field" is a long roll of Tyvek housewrap taped to the protective mats on the gym floor.

Each child has a personalized "parking spot" on which to place a rig.

We brought all our rigs to the gym and the parents and friends began to arrive.

Ships are placed on the personalized parking spots.

Everyone has a place of their own.

Guests arrive.

We are getting organized.

More and more ships.

The gym is starting to fill up.

More and more people came and the events began.
First the rib-rig race.

OK 4th grade.

Everyone wind up.

Get ready!

Get set!

Go!

There they go!

Now 5th grade.

Set down your rib – rigs
And wind them up.

Get ready!.

Get set.

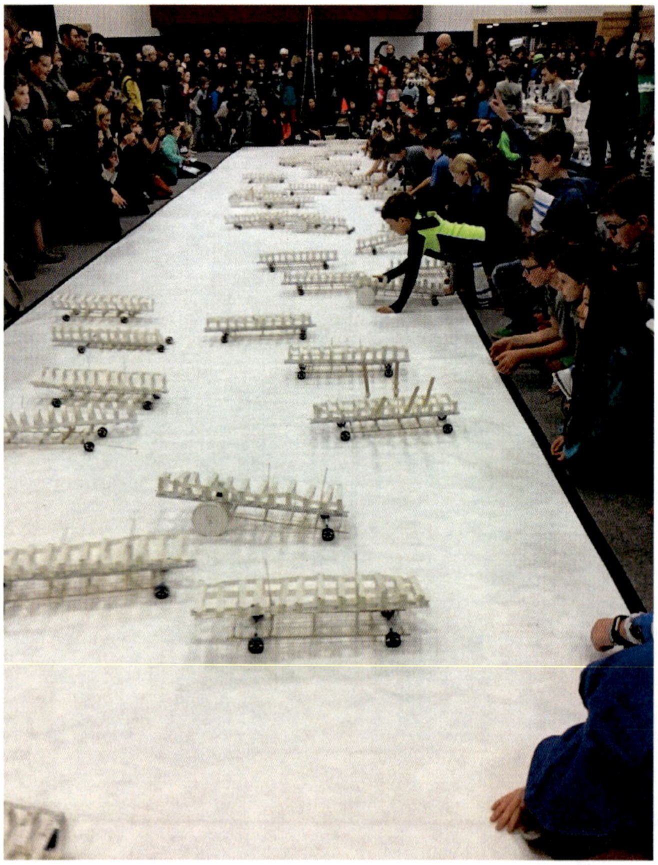

Go!

Then the launching of our tall ships. An armada!

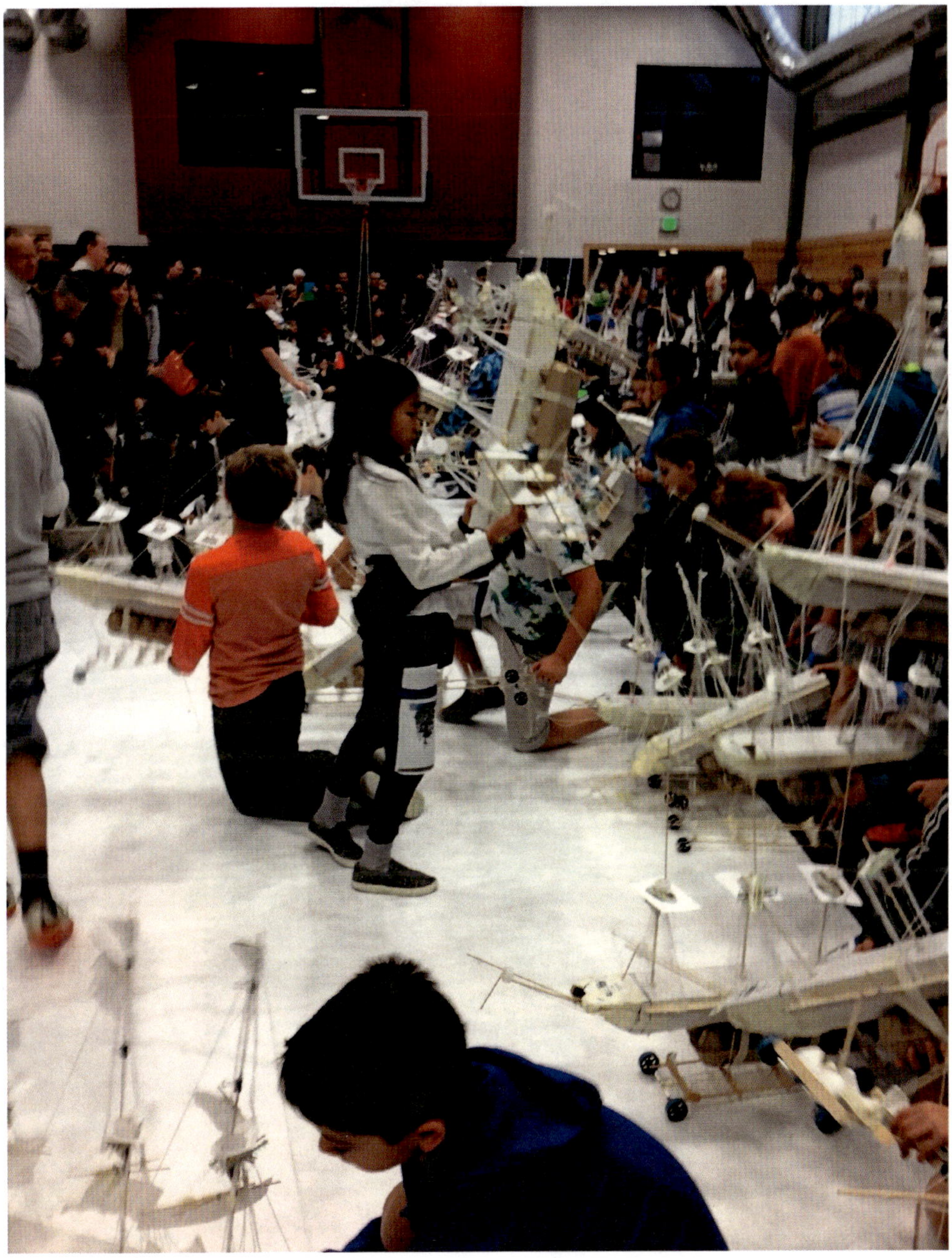

Oops.
Lots of them fell over.

Everyone had an entry form
taped to their arm, leg, or body.

We have to record how well
our rigs did in the gym.

More tall ships. 5th grade this time.

OK. That balanced a little better.

Finally it was time to pack up.

Each person had a big clear bag for the ship and a blue bag for rib-rigs and cradles. Take your treasures home carefully.

Everyone ripped off their entry form and ran to the multi-purpose room to trade them for a yummy-tummy prize.

What a lot of trophies to take home!

And wonderful memories to take home too!
Doc "O"

Building ships.
Building understanding.
Building children.

Made in United States
North Haven, CT
05 November 2026

21760281R00071